T0332605

Nonlinear Stochastic Systems Theory and Applications to Physics

Mathematics and Its Applications

George Adomian
Center for Applied Mathematics,
University of Georgia, Athens, Georgia, U.S.A.

Nonlinear Stochastic Systems Theory and Applications to Physics

KLUWER ACADEMIC PUBLISHERS
DORDRECHT / BOSTON / LONDON

Library of Congress Cataloging in Publication Data

Adomian, G.
Nonlinear stochastic systems theory and applications to physics / George Adomian.
p. cm.
Includes index.
ISBN 902772525X
1. Stochastic systems. 2. Nonlinear theories. I. Title.
QC20.7.S8A36 1988
530.1'5--dc19 88-29779
CIP

ISBN 90-277-2525-X

Published by Kluwer Academic Publishers,
P.O. Box 17, 3300 AA Dordrecht, The Netherlands.

Kluwer Academic Publishers incorporates
the publishing programmes of
D. Reidel, Martinus Nijhoff, Dr W. Junk and MTP Press.

Sold and distributed in the U.S.A. and Canada
by Kluwer Academic Publishers,
101 Philip Drive, Norwell, MA 02061, U.S.A.

In all other countries, sold and distributed
by Kluwer Academic Publishers Group,
P.O. Box 322, 3300 AH Dordrecht, The Netherlands.

Printed in The Netherlands

This book is dedicated to

Fred C. Davison and David S. Saxon

Series Editor's Preface

Approach your problems from the right end
and begin with the answers. Then one day,
perhaps you will find the final answer.

"The Hermit Clad in Crane Feathers" in R.
van Gulik's *The Chinese Maze Murders.*

It isn't that they can't see the solution.
It is that they can't see the problem.

G. K. Chesterton. *The Scandal of Father
Brown* . "The Point of a Pin."

Growing specialization and diversification have brought a host of monographs and textbooks on increasingly specialized topics. However, the "tree" of knowledge of mathematics and related fields does not grow only by putting forth new branches. It also happens, quite often in fact, that branches which were thought to be completely disparate are suddenly seen to be related.

Further, the kind and level of sophistication of mathematics applied in various sciences has changed drastically in recent years: measure theory is used (non-trivially) in regional and theoretical economics; algebraic geometry interacts with physics; the Minkowsky lemma, coding theory and the structure of water meet one another in packing and covering theory; quantum fields, crystal defects and mathematical programming profit from homotopy theory; Lie algebras are relevant to filtering; and prediction and electrical engineering can use Stein spaces. And in addition to this there are such new emerging subdisciplines as "experimental mathematics", "CFD", "completely integrable systems", "chaos, synergetics and large-scale order", which are almost impossible to fit into the existing classification schemes. They draw upon widely different sections of mathematics. This program, Mathematics and Its Applications , is devoted to new emerging (sub) disciplines and to such (new) interrelations as exempla gratia:

- a central concept which plays an important role in several different mathematical and/or scientific specialized areas;
- new applications of the results and ideas from one area of scientific endeavor into another;

- influences which the results, problems and concepts of one field of enquiry have and have had on the development of another.

The Mathematics and Its Applications program tries to make available a careful selection of books which fit the philosophy outlined above. With such books, which are stimulating rather than definitive, intriguing rather than encyclopaedic, we hope to contribute something toward better communication among the practitioners in diversified fields.

This particular book in the series is about solving equations: linear and nonlinear ordinary and partial differential equations, linear and nonlinear stochastic ordinary and partial differential equations. Here the word "solve" is used advisedly. It is definitely not a book on the theory of these things: it is, on the contrary, very much a book on how to get an answer, in terms of calculable numbers and functions, once a particular equation is given. The method exposed, in many forms and guises, and in terms of many examples and applications is the so-called "decomposition method". This is a method which appears to work well, occasionally very well in practice, but whose theoretical foundations still leave much to be explored (cf. below) and which thus poses a considerable and fascinating challenge. This is of course far from the only practically important method for solving problems whose justification is largely missing. Thus it appears to be routine in certain large-scale econometric-planning optimization problems to use algorithms designed and justified for convex problems even when the problem at hand very clearly is not convex: often with excellent results. One can object to such practices, which seem to be quite widespread, but the more interesting intellectual challenge is to find out why things often, or occasionally, go right anyway, and to find out when, i.e. in what circumstances, this is likely to happen. This, at least to a not inconsiderable extent, seems to be the situation with the decomposition method. It very often works well, but when and why, given that it is easy to write down in cases where it is clearly a very doubtful way to proceed, is a mystery.

Let me make that more precise. In one of its simpler guises, one writes the differential equation to be solved in the form

$$Lu + Ru + Nu = x \tag{1}$$

where L is a differential operator, say d^n/dt^n, which, using the initial conditions, is easy to invert, R is the remaining linear part, and

N is a nonlinear (analytic) operator, e.g. $Nu = u^2$ is a very simple case. Now imagine that u is to be found as a series $u = u_0 + u_1 + \cdots$. One determines u_0 from initial and for boundary conditions in some suitable way to obtain an equation with known u_0

$$u = u_0 + L^{-1}Ru - L^{-1}Nu$$

Expand Nu in some suitable way as a sum

$$Nu = \Sigma A_n(Nu)$$

where the $A_n(Nu)$ are only allowed to depend on $u_0, \ldots u_n$. There is choice here, which can be advantageously exploitèd. Now rewrite equation (1) as

$$u_0 + u_1 + u_2 + \cdots = u_0 - L^{-1}Ru_0 - L^{-1}Ru_1 - \cdots -$$
$$L^{-1}A_0(Nu) - L^{-1}A_1(Nu) - \cdots$$

and ensure a solution by requiring

$$u_1 = -L^{-1}Ru_0 - L^{-1}A_0(Nu), \ldots, u_{n+1} = -L^{-1}Ru_n - L^{-1}A_n(Nu), \cdots$$

In the case of zero initial conditions with $N = 0$, this simply amounts to a formal inversion of the operator $(I + L^{-1}R)$ as a von Neumann series

$$(I + L^{-1}R)^{-1} = I - L^{-1}R + L^{-1}RL^{-1}R - L^{-1}RL^{-1}R + \cdots$$

from which it is abundantly clear that the method cannot possibly work in all circumstances. It does work, though, for whole classes of equations; that is, there is convergence in practice. In the case $N \neq 0$, this is definitely different from writing the formal inverse of $I + L^{-1}(R + N)$ and thus the method is definitely not something like Picard iteration as has been claimed.

One systematic way of using the method is to insert a formal counting parameter λ in (1) to obtain $Lu + \lambda Ru + \lambda Nu = x$ and to write $u = \Sigma\lambda^n u_n$, $Nu = \Sigma\lambda^n A_n(Nu)$. For $Nu = u^2$, for example, this then gives $A_n(Nu) = \Sigma_{k+1=n}\, u_k u_1$. Inserting all this in the new equation and equating powers of λ then gives again $u_{n+1} = L^{-1}Ru_n - L^{-1}A_n(Nu)$.

Finally set $\lambda = 1$. In a certain way these are deformation theoretic ideas. By introducing λ (for λ near 1 intuitively) the various contributions to the solution can be separated (unravelled) to be put together again subsequently.

The basic idea extends straightforwardly to more complicated situations involving partial stochastic differential operators. There is, as has been said, choice in the A_n and thus there is also an aspect of art to using the method. It is clear that the sums Σu_n cannot possibly converge in all circumstances. On the other hand, there are results due to the author and to N. Bellomo which say that in certain circumstances there is convergence. There are also heuristic arguments that the method ought to work better as a rule than e.g. Picard iteration. It should be stressed that the method, when used well, often yields solutions to remarkable accuracy (in the sense that one obtains solutions which, when substituted, work). Tests on e.g. a Cyber 205 show this; e.g. for Duffing and van der Pol equations with stochastic parameters in them.

Thus we appear to have a powerful, useful, and practical method as will be clear from the many problems discussed in this book; its range of applicability is almost a total mystery. So, all in all, here is a challenge to both practitioners and theoreticians.

The unreasonable effectiveness of mathematics
in science....
Eugene Wigner

Well, if you know of a better 'ole, go to it.
Bruce Bairnsfather

What is now proved was once only imagined.
William Blake

As long as algebra and geometry proceeded
along separate paths, their advance was slow
and their applications limited. But when
these sciences joined company, they drew from
each other fresh vitality and thenceforward
marched on at a rapid pace toward perfection.
Joseph Louis Lagrange

Michiel Hazewinkel
Amsterdam
May, 1988

Table of Contents

Acknowledgments

I am deeply appreciative of the patient, cheerful, and dedicated work of Karin Haag who prepared the camera-ready manuscript for this book on a Macintosh computer. I would like to thank Diane Cronin and Arlette Revells, who typed preliminary drafts and papers on which this book is based. As always, I am grateful to my wife, Corinne Adomian, who edited this manuscript.

I am indebted to President Fred Davison, who provided the opportunity to do this work by instituting and supporting the Center for Applied Mathematics as part of his vision for The University of Georgia. His vision and support made our work possible and opened the door to much further work. Dr. James Kenney also has been a very special friend whose help was invaluable. I appreciate the support of numerous colleagues in the United States, England, Italy, The Netherlands, Austria, Japan, Soviet Armenia, and other countries. Finally, I would like to thank collaboraters Randolph Rach, Dr. McLowery Elrod, Dr. Nicola Bellomo, Dr. Ida Bonzani, Dr. R. Riganti and Dr. Gerald E. Adomian for suggestions and discussions on pertinent papers.

George Adomian

About the Typesetting of this Book

This manuscript was completed under difficult and trying circumstances. Obstacles were overcome, however, thanks to the help of many people. Dr. Harvey Lam of Princeton University created Princeton font, the mathematical program used for this manuscript. He made the program available to me and answered questions about it which was very helpful. Others who gave of their time and expertise were Steve Phillips, Jerry Freilich, Clate Sanders, Jeff Parker, and Bert DeSimone. Without their assistance a final copy could not have been produced.

I also wish to thank Diane Cronin, Arlette Revells, and Corinne Adomian for typing, editing, and proofreading assistance. Randolph Rach and Dr. McLowrey Elrod helped with proofing and provided invaluable support. Barry Wood's technical advice was appreciated. The author was also a tireless proofreading assistant.

Dr. Fred Davison made both time and computer facilities available so that this manuscript was completed. As both an editor and typesetter, I am especially grateful for his support.

Karin Haag

Author's Preface

This book has been written for the applied mathematician as well as for physicists, engineers, biologists, and economists interested in solving real, physical problems. It is not intended for abstract formulations and pathological cases. For pure mathematicians the work should be a valuable source for abstract reformulations.

Part I summarizes the decomposition method for the convenience of readers. The previous books, *Stochastic Systems* (Academic Press, 1983) and *Nonlinear Stochastic Operator Equations* (Academic Press, 1986), provide more detail and cover some topics not included here. Part II considers applications to various physical problems. We have shown how solutions can be obtained if initial/boundary conditions are specified which are realistic and appropriate to the problem. Ideally, each chapter should have ended with detailed numerical results and comparisons with usual methods. Unfortunately, no assistance or computational facilities were available, so tables, computations, and other rich resources are unavailable.

Useful criticism for continued work is appreciated, and it is accepted a priori that errors may exist which we will be happy to learn about. We believe the book will be useful to many (especially to those who read earlier books and papers). No bibliography on the background of the applications or their treatment by other methods is provided since this is readily available elsewhere.

The advantages for this methodology are avoidance of assumptions and restrictions which necessarily change the problem to a different problem to make it "tractable." A large number of examples in the earlier books and papers (as well as in unpublished work) have demonstrated accuracy and much decreased computation. A rigorous structure can now be profitably pursued by mathematicians while physicists study the applications. In the author's opinion, the wide range of applicability makes the work valuable to researchers and students alike. The question of validity of the decomposition solution is re-examined in the Epilogue.

This book is not intended to be the final word on the subject. On the contrary, it is intended as an introduction for new work both on the suggested approaches to applications and on a rigorous mathematical structure and understanding of the range of applicability, properties, and generation of the A_n polynomials, convergence, and other matters. Neither is it intended to be an encyclopedic reference to other work nor to explain the background or derivation of the well-known model equations considered.

Much research on now-standard methods is omitted for good reasons. Our objective is mathematics accessible to engineers, physicists, biologists and economists concerned with real-world quantitative solutions, not formal theory and deep theorems on systems mathematized for tractability with underlying assumptions which are nonphysical. The book deals with areas intellectually exciting and interesting to the author, with no attempt to explain or recapitulate work already in the literature on the background of the applications or other approaches.

I believe the method offers an approach with a potential for progress and accurate, efficient, quantitative results to many problems. It seems particularly adaptable to the new highly parallel computers. Hopefully it will be received with open minds in the spirit of Francis Bacon, "Read not to contradict or confute, nor to believe and take for granted, but to weigh and consider."

George Adomian
Athens, Georgia

Part I

A Summary of the Decomposition Method

Introduction: The next three chapters explore methodology for solution of nonlinear equations, the A_n polynomials for representation of nonlinear functions, effects of linearization, and discussion of initial/boundary conditions. In Part II applications are considered.

CHAPTER 1

The Decomposition Method

1.1 Introduction

Solution of linear and nonlinear stochastic equations can be carried out using the decomposition method (1983; 1986). Assuming the general form $Fu = x(t)$, an operator equation, the solution is assumed to be a decomposition into components $\sum_{n=0}^{\infty} u_n(t)$ where u_0 can be determined from $x(t)$, the initial or boundary conditions, and the invertible part of the linear deterministic component of F. An additional step for nonlinear equations is the representation of the nonlinear component Nu by $\sum_{n=0}^{\infty} A_n(u_0, u_1, \ldots u_n)$ where the A_n are polynomials defined by the author and generated for the specific nonlinearity. Although as shown in (1986), we can consider nonlinear terms of the form $g(u,u',\ldots)$ or even composite nonlinearities and radicals, let us confine our attention now to $Nu = f(u)$ where $f(u)$ is an analytic function.

Just as $f(u)$ can be decomposed into a Fourier series of sines and cosines, we have seen that $f(u)$ can be decomposed into a rapidly convergent set of polynomials A_n where $A_n = A_n(u_0, u_1, \ldots, u_n)$. This is the key factor in being able to make the decomposition solution without perturbation or smallness assumptions, closure approximations, etc. When any nonlinear term is expressed in terms of the A_n, then the decomposition of u into components u_n and the identification of u_0 allows all following terms to be computed so an n-term approximation $\phi_n = \sum_{\nu=0}^{n-1} u_\nu$, for some n, will then represent u to the desired accuracy.

In the previous work, we have pointed out that the A_n are not uniquely defined for any $f(u)$. This is by no means a disadvantage, and finding still other formulations of these polynomials in convenient forms becomes a rich field of research. Let us first summarize the concept of the decomposition method for completeness in this book; then we will review generation schemes for the A_n; and we finally present another version of the polynomials which offers some important advantages.

1.2 Summary of the Decomposition Method

Consider an equation in the form $Lu + Nu = x(t)$ where L is a linear operator and N represents a nonlinear operator, with L invertible under sufficient existence and regularity conditions in the pertinent function space so that L^{-1} exists and the functions $f(x)$ on which L^{-1} acts are such that $L^{-1}f(x)$ is measurable. In this case, we can write $Lu = x(t) - Nu$ and consequently $u(x,t) = u_0 - L^{-1}Nu$ where $u_0 = \Phi + L^{-1}x$ with Φ satisfying $Lu = 0$. Preferably, we will write an equation with a linear part and a nonlinear part as $Lu + Ru + Nu = x(t)$ where L is easily or trivially invertible and let R be the remaining linear part. (For an nth-order differential equation we let $L = d^n/dt^n$ and L^{-1} be an n-fold integration from 0 to t in an initial condition problem.) Then $Lu = x(t) - Ru - Nu$, and we proceed as before. In cases of differential equations with stochastic process coefficients $a_\upsilon(t,\omega)$ and stochastic process $x(t)$, i.e., $x(t,\omega)$, $t \varepsilon T$, $\omega \varepsilon (\Omega, F, \mu)$, a probability space, we consider $\mathcal{L} = \sum_{\upsilon=0}^{n} a_\upsilon(t,\omega) d^\upsilon/dt^\upsilon$ with a_n defined as 1 and write $L = \sum_{\upsilon=0}^{n} \langle a_\upsilon \rangle d^\upsilon/dt^\upsilon$ where $\langle a_\upsilon \rangle$ is an expectation and $R = \sum_{\upsilon=0}^{n-1} \alpha_\upsilon(t,\omega) d^\upsilon/dt^\upsilon$, with $\alpha_\upsilon = a_\upsilon - \langle a_\upsilon \rangle$ is a random operator. Again, more conveniently, we write $Lu + Ru + \mathcal{R}u + Nu = x(t)$ where $L = d^n/dt^n$, let R represent the remaining linear deterministic terms, and $\mathcal{R}u$ the linear part involving stochastic processes. The nonlinear term can also be written as $\mathcal{N}u = Nu + \mathcal{M}u$ where Nu is a deterministic term and $\mathcal{M}u$ is a stochastic term. (The latter two will be similarly treated using the A_n polynomials.) Now the decomposition $u = \sum_{n=0}^{\infty} u_n$ and the fact that $A_n = A_n(u_0, u_1, \ldots, u_n)$ with Nu and $\mathcal{M}u$ written in infinite sums of A_n polynomials results in a series

$$\sum_{n=0}^{\infty} u_n(t) = u_0 - L^{-1}R \sum_{n=0}^{\infty} u_n - L^{-1}\mathcal{R} \sum_{n=0}^{\infty} u_n - L^{-1} \sum_{n=0}^{\infty} A_n(Nu) - L^{-1} \sum_{n=0}^{\infty} A_n(\mathcal{M}u)$$

where the notation $A_n(Nu)$ means the A_n are generated for the term Nu.

Now identifying terms in the summation, we identify u_1

with $- L^{-1}Ru_0 - L^{-1}\mathcal{R}\,u_0 - L^{-1}A_0(Nu) - L^{-1}A_0(\mathcal{M}\,u)$. Thus u_1 is calculable in terms of the known u_0. Proceeding in this way, we have

$$u_{n+1} = - L^{-1}Ru_n - L^{-1}\mathcal{R}\,u_n - L^{-1}A_n(Nu) - L^{-1}A_n(\mathcal{M}\,u)$$

so all components of u are calculable and $u = \sum_{n=0}^{\infty} u_n$. The generally rapid convergence is discussed later as well as in the referenced works. For deterministic equations, $\mathcal{R}\,u$ and $\mathcal{M}\,u$ vanish, and we have an ordinary series. When they do not, we have a stochastic series from which first-and second-order statistics can be obtained since an n-term approximation $\phi_n = \sum_{\upsilon=0}^{n-1} u_\upsilon$ for low n generally suffices.

The use of $L = d^n/dt^n$ avoids the problem of difficult Green's functions and results in simplified integrations.

When we deal with partial differential equations involving $u(x,y,z,t)$, we may have linear terms $L_t u$, $L_x u$, $L_y u$, $L_z u$, and we must solve for each linear term in turn, do the inversion, add the resulting (in this case four) equations and divide by the number of equations. We have $u = \sum_{n=0}^{\infty} u_n(x,y,z,t)$ and use the A_n as before, thus solving as if we have an ordinary differential equation. The treatment is easily extended to the treatment of systems of differential and partial differential equations. For n coupled differential equations, we simply obtain a set of n u_0 components, in terms of which the set of n u_1 components is computable, etc. A system of n partial differential equations uses both the latter ideas, and we find the decomposition now applying on a wide scale and linear and/or deterministic cases becoming special cases of a general treatment. While a mathematically prepared reader can continue, the reading of the references will answer many questions.

Example:

Tsunami propagation: This problem was previously solved by George F. Carrier. Using a linear stochastic Mathieu equation model,

$$d^2u(x)/dx^2 + [1 + \varepsilon\, f(x)]u(x) = 0$$

with random $f(x)$, and $u(0) = 1$, $u'(0) = 0$. Let $L = d^2/dx^2$ and let L^{-1} denote the two-fold integral operator. Thus

$$Lu = -[1 + \varepsilon f(x)]u$$

$$L^{-1}Lu = -L^{-1}[1 + \varepsilon f(x)]u$$

$$u - u(0) - xu'(0) = -L^{-1}[1 + \varepsilon f(x)]u$$

or

$$u = u_0 - L^{-1}[1 + \varepsilon f(x)] \sum_{n=0}^{\infty} u_n$$

where we let $u_0 = u(0) + xu'(0)$ and $u = \sum_{n=0}^{\infty} u_n$. Now

$$u_0 = 1$$

$$u_1 = -L^{-1}[1 + \varepsilon f(x)]u_0$$

$$u_2 = -L^{-1}[1 + \varepsilon f(x)]u_1$$

$$u_3 = -L^{-1}[1 + \varepsilon f(x)]u_2$$

$$\vdots$$

where the u_n for $n \geq 1$ are evaluated with an assumed $f(x)$. If we use only a two-term approximation for u, i.e., $u = u_0 + u_1$, assuming $\langle f(x)\rangle = 0$ and $\langle f(x)f(x')\rangle = \delta(x - x')$, we have

$$\langle u^2\rangle = 1 - x^2 + x^4/4 + \varepsilon^2 x^3/3!$$

as an approximation to the mean square fluctuation in height. We can now plot this as a function of x with ε as a parameter. The results must conform to Professor Carrier's original results for small ε. Noting that we have used no closure approximations or perturbation theory, ε need not be small and these results are equally valid for $\varepsilon = 1$ when perturbation fails. We can also drop the assumption of a delta-correlated process made here for convenient comparison. Such processes are assumed for mathematical convenience and are non-physical. The decomposition method here is not limited to linearity, harmonic waves (quasi-monochromaticity), Wiener processes, etc. and can treat a more sophisticated model where u depends on x,y,z,t and is described by a nonlinear stochastic equation.

Example:

As a simple (linear deterministic) example, consider the Airy's equation $y'' - ty = 0$; $y(0) = 1$; $y'(0) = 1$ written in the form $Ly - Ry = 0$ with $L = d^2/dt^2$, $R = t$, L^{-1} defined as a two-fold definite integration from 0 to t. Operating with L^{-1} we obtain $y(t) = y(0) + ty'(0) + L^{-1}Ry$. Thus

$$y_0 = 1 + t$$

$$y_1 = L^{-1}Ry_0 = L^{-1}t(1 + t) = \frac{t^3}{2\cdot 3} + \frac{t^4}{3\cdot 4} + \frac{1\cdot t^3}{3!} + \frac{2\cdot t^4}{4!}$$

$$y_2 = L^{-1}Ry_1 = \frac{t^6}{2\cdot 3\cdot 5\cdot 6} + \frac{t^7}{3\cdot 4\cdot 6\cdot 7} + \frac{1\cdot 4\cdot 6\cdot t^6}{6!} + \frac{2\cdot 5\cdot t^7}{7!}$$

$$\vdots$$

$$y_n = \frac{1\cdot 4\cdot 7 \ldots (3n-2)t^{3n}}{(3n)!} + \frac{2\cdot 5\cdot 8 \ldots (3n-1)t^{3n+1}}{(3n+1)!}$$

Example:

Suppose now we consider the equation $d^2u/dx^2 - 40xu = 2$ with $u(-1) = u(1) = 0$. This is the one-dimensional case of the elliptic equation $\nabla^2 u = f(x,y,z) + k(x,y,z)u$ arising in problems of physics and engineering. Here $L = d^2/dx^2$, and we have $Lu = 2 + 40\,xu$. This is a relatively stiff case because of the large coefficient of u, and the nonzero forcing function yields an additional Airy-like function. Here as in general, L^{-1} involves indefinite integrations (the definite integrals are merely more convenient in initial value problems. Operating with L^{-1} yields $u = A + Bx + L^{-1}(2) + L^{-1}(40xu)$. Let $u_0 = A + Bx + L^{-1}(2) = A + Bx + x^2$ and let $u = \sum_{n=0}^{\infty} u_n$ with the components to be determined so that the sum is u. We identify $u_{n+1} = L^{-1}(40xu_n)$. Then all components can be determined, e.g., $u_1 = (20/3)Ax^3 + (10/3)Bx^4 + 2x^5$ and $u_2 = (80/9)Ax^6 + (200/63)Bx^7 + (10/7)x^8$, etc. Now we can calculate an n-term approximant $\phi_n = \sum_{i=0}^{n-1} u_i$. We will use $n = 12$ for numerical results. For $x = 0.2$, we get -0.135649. For $x = 0.4$, we get -0.113969. For $x = 0.6$, we get -0.083321. For $x = 0.8$, we get -0.050944. For $x = 1.0$, of course, we get zero. These easily obtained results are correct to seven digits. This is discussed in greater detail in Chapter 2. We see that a better solution is obtained and much more easily than by variational methods. The solution is found just as easily for nonlinear versions without linearization.

1.3 Generation of the A_n Polynomials

In order to calculate the A_n as in earlier works, the decomposition of the solution $u(t)$ into components to be determined via the expression $u = \sum_{n=0}^{\infty} u_n(t)$ is parametrized into $u = \sum_{n=0}^{\infty} \lambda^n u_n(t)$. It is necessary to emphasize again that λ is *not* a "small" parameter.[1]

Suppose as before that the nonlinear term $Nu = f(u)$ is now represented as $\sum_{n=0}^{\infty} \lambda^n A_n$. Now $f(u)$, or $f(u(\lambda))$, is assumed to be analytic in λ. The A_n are polynomials defined such that each A_n depends only on $u_0, u_1, \ldots, u_n$, i.e., only on components from u_0 to u_n. Thus $A_0 = A_0(u_0)$, $A_1 = A_1(u_0, u_1)$, $A_2 = A_2(u_0, u_1, u_2)$, etc. Then, in the decomposition series which results, each component u_n is calculable from the preceding term u_{n-1}. The u_0 term is explicitly known. It is obtained from the input or forcing term, the initial or boundary conditions, and the invertible part of the linear operator.

The latter is an important point not always understood. We do not invert the entire linear operator; consequently, we obtain simpler Green's functions than we would otherwise. This makes calculation of the resulting integrals much easier. The u_{n+1} term depends on u_n and the highest term in A_n is u_n; therefore, the system is calculable without perturbation or closure approximation. Now we must see how the A_n can be obtained. Once they are, they can be viewed as a special convenient set of polynomials like Hermite's

---

[1] In the new scheme which follows, the parametrization is avoided and $u = \sum_{n=0}^{\infty} u_n(t)$ and $Nu = \sum_{n=0}^{\infty} \hat{A}_n$ instead of $\sum_{n=0}^{\infty} A_n$ as before. The $\hat{A}_n$ polynomials should not be confused with the A_n.

polynomials or Legendre's polynomials and can become even more familiar because of the global nature of the methodology as we will see in the following chapters. They can be calculated from the following expression:

$$A_n = (1/n!)(d^n/d\lambda^n)f(u(\lambda))\big|_{\lambda=0} \qquad (1.2.1)$$

If we write $D = d/d\lambda$, we can write (1.2.1)

$$A_n = (1/n!)\, D^n f\big|_{\lambda=0}$$

We have now a systematic scheme with $D = d/d\lambda = (du/d\lambda)(d/du)$ since $f = f(u)$ and $u = u(\lambda)$. Each $D^n f$ is evaluated at $\lambda = 0$ and divided by $n!$ Since $u = u_0 + \lambda u_1 + \lambda^2 u_2 + \cdots$, the following are useful relations:

$$(d^n/d\lambda^n)u(\lambda)\big|_{\lambda=0} = n!\, u_n$$

$$(d^n/du^n)f(u(\lambda))\big|_{\lambda=0} = d^n f/du_n = h_n(u_0) \qquad (1.2.2)$$

(Since $d^n f/du^n$ is a function of u_0, we have written it as $h_n(u_0)$.)

The $D^n f$ for $n > 0$ can be written as a sum from $\upsilon = 1$ to n of terms $d^\upsilon f/du^\upsilon$ with coefficients which are polynomials in the $d^\upsilon u/d\lambda^\upsilon$. Thus,

$$D^1 f = (\partial f/\partial u)(\partial u/\partial\lambda)$$

$$D^2 f = (\partial^2 f/\partial u^2)(\partial u/\partial\lambda)^2 + (\partial f/fu)(\partial^2 u/\partial\lambda^2) \qquad (1.2.3)$$

$$D^3 f = (\partial^3 f/\partial u^3)(\partial u/\partial\lambda)^3 + 3(\partial^2 f/\partial u^2)(\partial u/\partial\lambda)(\partial^2 u/\partial\lambda^2)$$
$$+ (\partial f/\partial u)(\partial^3 u/\partial\lambda^3)$$

.
.
.

If for the nth derivative $D^n f$, we denote the υth coefficient by $c(\upsilon,n)$, we can write

$$D^n f = \sum_{\upsilon=1}^{n} c(\upsilon,n)F(\upsilon) \qquad (1.2.4)$$

where $F(\upsilon) = d^\upsilon f/du^\upsilon$. D^3f, for example, is given by $D^3f = c(1,3)F(1) + c(2,3)F(2) + c(3,3)F(3)$. The second index in the $c(\upsilon,n)$ is the order of the derivative and the first index progresses from 1 to n along with the index of F.

These coefficients can be calculated in a number of ways. The first calculations were done by developing a recurrence relation which can be given as follows: For $1 \leq i,\ j \leq n$,

$$c(i,j) = (d/d\lambda)\{c(i,j-1)\} + (dy/d\lambda)\{c(i-1,j-1)\} \qquad (1.2.5)$$

with $c(0,0) = 1$ and $c(1,0) = 0$. The second is true because $c(i,j) = 0$ for $i > j$. That the coefficient $c(0,0) = 1$ is seen by comparison to $D^0f = c(0,0)d^0f/du^0 = f$, i.e., $A_0 = f|_{\lambda=0} = f(u_0)$.

The notation

$$\psi(i,j) = (d^iu/d\lambda^i)^j \qquad (1.2.6)$$

$$F(i) = d^if/du^i$$

will be convenient because these quantities are explicit derivatives, and the implicit differentiations of (1.1.1) are cumbersome.

Now (1.1.4) can be given in terms of the c's or the ψ's, for example: $D^3f = c(1,3)F(1) + c(2,3)F(2) + c(3,3)F(3)$ as above, or

$$D^3f = \psi(3,1)F(1) + 3\psi(1,1)\psi(2,1)F(2) + \psi(1,3)F(3)$$

$$= (\partial^3u/\partial\lambda^3)(\partial f/\partial u) + 3(\partial u/\partial\lambda)(\partial^2u/\partial\lambda^2)(\partial^2f/\partial u^2)$$

$$+ (\partial u/\partial\lambda)^3(\partial^3f/\partial u^3)$$

While we can now derive the A_n quickly or simply present generating forms, we will go into detail for the benefit of mathematical researchers or persons seeking dissertation topics. Let us consider the $c(i,j)$ coefficients.

$$c(0,0) = \psi(1,0) = 1$$

$$c(1,1) = (d/d\lambda)\{c(1,0)\} + \psi(1,1)\{c(0,0)\}$$

$$= (d/d\lambda)\ \{0\} + \psi(1,1)\{1\} = \psi(1,1)$$

Noting that $c(0,j) = 0$ for $j > 0$

$$c(2,2) = d/d\lambda\{c(2,1)\} + \psi(1,1)\{c(1,1)\}$$

$$= d/d\lambda\{0\} + \psi(1,1)\ \psi(1,1) = \psi(1,2)$$

$$c(1,2) = d/d\lambda\{c(1,1)\} + \psi(1,1)\{c(0,1)\}$$

$$= d/d\lambda\{\psi(1,1) + \psi(1,1)\{0\} = \psi(2,1)$$

$$c(3,3) = d/d\lambda\{c(3,2)\} + \psi(1,1)\{c(2,2)\}$$

$$= \psi(1,1\ \psi(1,2) = \psi(1,3)$$

which are written in detail for the convenience of students.

$$c(2,3) = d/d\lambda\ \{c(2,2)\} + \psi(1,1)\{c(1,2)\}$$

$$= d/d\lambda\ \psi(1,2) + \psi(1,1)\psi(2,1)$$

$$= 2\psi(1,1)\psi(2,1) + \psi(1,1)\ \psi(2,1)$$

$$= 3\psi(1,1)\ \psi(2,1)$$

$$c(1,3) = d/d\lambda\ \{c(1,2)\} = d/d\lambda\ \psi(2,1) = \psi(3,1)$$

To calculate any $D^n f$, one needs only to multiply the $c(\upsilon,n)$ by $F(\upsilon)$ for the range from 1 to n and to calculate the sum. The A_n are then given by:

$$A_n = (1/n!)D^n f \,|_{\lambda=0}$$

Thus, remembering the definition from (1983) that $h_n(u_0) = (d^n/du^n)f(u(\lambda))\,|_{\lambda=0}$

$$A_0 = (1/0!)D^0 f\,|_{\lambda=0} = f(u_0) = h_0(u_0)$$

$$A_1 = (1/1!)D^1 f\big|_{\lambda=0}$$

$$= c(1,1)h_1(u_0) = \psi(1,1)h_1(u_0) = h_1(u_0)u_1$$

$$A_2 = (1/2!)D^2 f\big|_{\lambda=0}$$

$$= (1/2)\{c(1,2)h_1 + c(2,2)h_2\}$$

$$= (1/2)\{\psi(2,1)h_1 + \psi(1,2)h_2\}$$

$$= (1/2)\{2u_2 h_1 + u_1^2 h_2\}$$

$$= (1/2)\{h_2(u_0)u_1^2 + 2h_1(u_0)u_2\}$$

$$= h_1(u_0)u_2 + h_2(u_0)(1/2)u_1^2$$

showing various arrangements for convenience. Continuing in this manner,

$$A_3 = (1/3!)\{c(1,3)h_1 + c(2,3)h_2 + c(3,3)h_3\}$$

$$= (1/3!)\{\psi(3,1)h_1 + 3\psi(1,1)\psi(2,1)h_2 + \psi(1,3)h_3\}$$

$$= (1/3!)\{(3!)u_3 h_1 + (3!)u_1 u_2 h_2 + u_1^3 h_3\}$$

$$= (1/3!)\{h_3(u_0)u_1^3 + 6h_2(u_0)u_1 u_2 + 6h_1(u_0)u_3\}$$

$$= h_1(u_0)u_3 + h_2(u_0)u_1 u_2 + h_3(u_0)(1/6)u_1^3$$

$$A_4 = (1/4!)\{h_4(u_0)u_1^4 + 12h_3(u_0)u_1^2 u_2$$

$$+ h_2(u_0)[12u_2^2 + 24u_1 u_3] + 24h_1(u_0)u_4\}$$

$$= h_1(u_0)u_4 + h_2(u_0)[(1/2)u_2^2 + u_1 u_3]$$

$$+ h_3(u_0)[(1/2)u_1^3 u_2] + h_4(u_0)(1/24)u_1^5$$

$$A_5 = h_1(u_0)u_5 + h_2(u_0)[u_2 u_3 + u_1 u_4]$$

$$+ h_3(u_0)[(1/2)u_1 u_2^2 + (1/2)u_1^2 u_3]$$

$$+ h_4(u_0)(1/6)u_1^3u_2 + h_5(u_0)(1/120)u_1^5$$

$$A_6 = h_1(u_0)u_6 + h_2(u_0)[(1/2)u_3^2 + u_2u_4 + u_1u_5]$$
$$+ h_3(u_0)[(1/6)u_2^3 + u_1u_2u_3 + (1/2)u_1^2u_4]$$
$$+ h_4(u_0)[(1/4)u_1^2u_2^2 + (1/6)u_1^3u_3]$$
$$+ h_5(u_0)(1/24)u_1^4u_2 + h_6(u_0)(1/720)u_1^6$$

We observe that in the linear case, i.e., $f(u) = u$, we have $A_n \equiv u_n$, and the results will conform to previously derived results for linear equations in (1983). Thus $f(u) = u$ results in $h_0 = u_1$, $h_1 = 1$, and $h_i = 0$ for $i \geq 2$; as a consequence, $A_0 = u_0$, $A_1 = u_1$, ..., $A_n = u_n$.

Alternatively, since $u = \sum_{n=0}^{\infty} \lambda^n u_n$ and $n!u_n = d^n u/d\lambda^n \big|_{\lambda=0}$

$$A_n = (1/n!)D^n f(u)\big|_{\lambda=0}$$
$$= (1/n!)D^n u\big|_{\lambda=0} = (1/n!)d^n u/d\lambda^n\big|_{\lambda=0}$$
$$= (1/n!)\{n!u_n\} = u_n$$

For the equation $Lu + Nu = x$, where L is a linear operator, consider the linear limit of Nu, i.e., $Nu = u$. We now have

$$Lu + u = x$$

$$Lu = x - u$$

$$u = u_0 - L^{-1}\Sigma A_n = u_0 + u_1 + u_2 + \ldots$$

Thus $u_{n+1} = u_{n+1}(u_n)$ so that in the linear case, u_{n+1} depends only on the term preceding it. In the nonlinear case, $u_{n+1} = u_{n+1}(u_0, u_1, \ldots u_n)$.

In addition to these simple forms of $f(u)$ which do not contain derivatives of u, e.g., u^2, u^4, e^u, $\sin u, \cdots$, we can also consider such terms as $f(u,u')$, $f(u,u',u'')$, etc. An example is:

$$Nu = [(d^0/dt^0)u][(d^1/dt^1)u] = uu' = f(u,u')$$

These are considered in the next section.

1.4 The A_n for Differential Nonlinear Operators

Consider the nonlinear operator $Nu = f(u,u^{(1)},\ldots,u^{(n)})$. We assume f is analytic. Then f is also analytic in λ, $u^{(0)},\ldots$. We are concerned with two important subcases of the *differential nonlinear operator* N which are:

1) sum of nonlinear functions of the time derivatives of u, with each nonlinear function dependent on a single derivative:

$$Nu = \sum_{i=0}^{n} N_i u = \sum_{i=0}^{n} f_i(u^{(i)})$$

2) a sum of products of nonlinear functions of u, each dependent on a single derivative. As an example, consider $f(u,u') = u^2(u')^3$.

Obviously, if $Nu = f(u)$, we have the simple nonlinearity for which we have previously found expansion coefficients (1983; 1986), and we must obtain identical results for this limiting case. We, therefore, defined the A_n for the general differential nonlinear operator Nu as

$$A_m = (1/m!)D^m\{f(u,u',\ldots,u^{(n)})\}\big|_{\lambda=0}$$

where $u,u',\ldots, u^{(n)}$ are assumed to be analytic functions of λ.

Case 1: The first subcase of our general class was specified by $Nu = f(u,u'\ldots,u^{(n)}) = \sum_{i=0} f_i(u^{(i)})$ which we will call a sum decomposition. The A_n are given by:

$$A_m = \sum_{i=0}^{n} [(1/m!)D^m f_i(u^{(i)})\big|_{\lambda=0}] = \sum_{i=0}^{n} A_{im}$$

because $f(u,u',\ldots,u^{(n)}) = \sum_{m=0}^{\infty} \lambda^m A_m = \sum_{i=0}^{n} f_i(u^{(i)})$ and each $f_i(u^{(i)}) = \sum_{m=0}^{\infty} \lambda^m A_{im}$. This leads to $f(u,u',\ldots,u^{(n)}) = \sum_{i=0}^{n} \sum_{m=0}^{\infty} \lambda^m A_{im} = \sum_{m=0}^{\infty} [\sum_{i=0}^{n} A_{im}]\lambda^m$ therefore $A_m = \sum_{i=0}^{n} A_{im}$.

Case 2: The second subcase, product decomposition of the nonlinear operator, decomposes $Nu = f(u,u',\ldots,u^{(n)})$ into a sum of products. Let us first take pairwise products such as $f(u,u') = f_0(u)f_1(u') = \prod_{i=0} f_i(u^{(i)})$ so we can consider nonlinearities such as u^2, $(u')^3$, etc. Now the A_n are given by:

$$A_m = (1/m!)D^m\{f_0(u^{(0)}(\lambda))f_1(u^{(1)}(\lambda))\}\big|_{\lambda=0}$$

$$= (1/m!)\sum_{k=0}^{m} \binom{m}{k}[D^{m-k} f_0(u^{(0)}(\lambda))]\,[D^k f_1(u^{(1)}(\lambda))]\big|_{\lambda=0}$$

This leads to

$$A_m = \sum_{k=0}^{m} A_{0,m-k}\; A_{1,k}$$

since

$$f(u,u') = \sum_{m=0}^{\infty} A_m \lambda^m = \prod_{i=0}^{1} f_i(u^{(i)})$$

and as before $f_i(u^{(i)}(\lambda)) = \sum_{m=0}^{\infty} A_{im}\lambda^m$. This implies $f(u,u') = (\sum_{m=0}^{\infty} A_{0m}\lambda^m)(\sum_{m=0}^{\infty} A_{im}\lambda^m) = \sum_{m=0}^{\infty}[\sum_{k=0}^{m} A_{0,m-k}A_{1k}]\lambda^m$ which also gives us the result. An extended Leibnitz rule in terms of multinomial coefficients can handle products of n factors, that is, $f(u,u',\ldots,u^{(n)}) = \prod_{i=0}^{n} f_i(u^{(i)})$.

1.5 Convenient Computational Forms for the A_n Polynomials

It is possible to find simple symmetry rules for writing the A_n quickly to high orders. Using the A_n, there is no need for mathematically inadequate and physically unrealistic approximations or linearizations.

Thus if the modeling retains the inherent nonlinearities, we may expect solutions conforming much more closely to actual behavior. We will consider here simple nonlinear operators not involving differentials, i.e., of the form $Nu = f(u)$. In the preceding section, we gave $A_n = (1/n!) \sum_{\upsilon=1}^{n} c(\upsilon,n)h_\upsilon(u_0)$ with $h_\upsilon(u_0) = (d^\upsilon/du^\upsilon)f(u(\lambda))|_{\lambda=0}$ with the $c(\upsilon,n)$ specified by a recurrence rule.

The A_n for polynomial nonlinearities being sums of various products of the u_i up to $i = n$ can also be written in symmetrized form. For $Nu = u^2 = \sum_{n=0}^{\infty} A_n$, for example, $A_0 = u_0^2$, $A_1 = 2u_0u_1$, $A_2 = u_1^2 + 2u_0u_2$, etc., but we can write this as $A_0 = u_0u_0$, $A_1 = u_0u_1 + u_1u_0$, $A_2 = u_0u_2 + u_1u_1 + u_2u_0$, etc., i.e., the first subscript goes from 0 to n, the second from n to 0 such that the sum is n.

In extending this to the forms we have just developed for the A_n, we begin by noticing that in order to determine $h_\upsilon(u_0)$ for $\upsilon = 1,2,\ldots,n$, we differentiate $f(u)$ υ times with respect to u and evaluate at $\lambda = 0$. Then, for example, A_3 would involve h_1, h_2, h_3. From the previous section $n!\, A_3 = c(1,3)h_1 + c(2,3)h_2 + c(3,3)h_3$. To get the $c(\upsilon,n)$ we simply ask how many combinations (not permutations) of υ integers will add to n.

Thus $c(\upsilon,n)$ will mean the product of υ u_i's whose subscripts add to n. To get $c(2,3)$, we see that two integers can add to 3 only if one is 1 and the other is 2, if zero is excluded. Hence, we write $c(2,3) = u_1u_2$. To get $c(1,3)$, the coefficient of $h_1(u_0)$, we have one u_i and its subscript must be 3, hence $c(1,3) = u_3^2$. $c(1,3)$ was previously $3!u_3$. The definition by this rule absorbs the n! Thus the present $c(\upsilon.n)$ includes the 1/n! What about $c(3,3)$, the coefficient of $h_3(u_0)$? Now we need three factors u_i with subscripts summing to 3; hence each subscript must be 1 and $c(3,3) = u_1u_1u_1 = u_1^3$. We divide by the factorial of the number of repetitions. We have the desired result, i.e., a simple heuristic rule to write down the $c(i,j)$ [2]. Then $c(3,3) = (1/3!)u_1$. We have now

$$A_3 = h_1(u_0)u_3 + h_2(u_0)u_1u_2 + h_3(u_0)(1/3!)u_1^3$$

For example, writing A_6, we need the coefficients for the terms $h_\upsilon(u_0)$ for υ from 1 to 6. The coefficient of h_6 must involve 6 integers adding to 6 or u_1^6; hence the coefficient of $h_6(u_0)$ is $(1/6!)u_1^6$. What about the coefficient for $h_2(u_0)$ in A_6 or $\upsilon = 2$, $n = 6$? Clearly we need two integers which sum to 6. These are

[2]This was determined in a number of conversations with R. Rach so we shall call it *Rach's Rule* .

(1,5), (2,4), and (3,3). Thus the coefficient $c(2,6)$ is $(1/2!)u_3^2 + u_2u_4 + u_1u_5$. The terms involve $\prod_{i=1}^{\nu} u_{k_i}$ with $\sum_{i=1}^{n} k_i = n$, and if we have j repeated subscripts, we divide by $j!$

$$A_0 = h_0(u_0)$$

$$A_1 = h_1(u_0)u_1$$

$$A_2 = h_1(u_0)u_2 + h_2(u_0)(1/2!)u_1^2$$

$$A_3 = h_1(u_0)u_3 + h_2(u_0)u_1u_2 + h_3(u_0)(1/3!)u_1^3$$

$$A_4 = h_1(u_0)u_4 + h_2(u_0)[(1/2!)u_2^2 + u_1u_3] + h_3(u_0)(1/2!)u_1^2u_2 + h_4(u_0)(1/4!)u_1^4$$

$$A_5 = h_1(u_0)u_5 + h_2(u_0)[u_2u_3 + u_1u_4] + h_3(u_0)[u_1(1/2!)u_2^2 + (1/2!)u_1^2u_3] + h_4(u_0)(1/3!)u_1^3u_2 + h_5(u_0)(1/5!)u_1^5$$

$$A_6 = h_1(u_0)u_6 + h_2(u_0)[(1/2!)u_3^2 + u_2u_4 + u_1u_5] + h_3(u_0)[(1/3!)u_2^3 + u_1u_2u_3 + (1/2!)u_1^2u_4] + h_4(u_0)[(1/2!)u_1^2(1/2!)u_2^2 + (1/3!)u_1^3u_3] + h_5(u_0)(1/4!)u_1^4u_2 + h_6(u_0)(1/6!)u_1^6$$

$$A_7 = h_1(u_0)u_7 + h_2(u_0)[u_3u_4 + u_2u_5 + u_1u_6] + h_3(u_0)[(1/2!)u_2^2u_3 + u_1(1/2!)u_3^2 + u_1u_2u_4 + (1/2!)u_1^2u_5] + h_4(u_0)[u_1(1/3!)u_2^3 + (1/2!)u_1^2u_2u_3 + (1/3!)u_1^3u_4] + h_5(u_0)[(1/3!)u_1^3(1/2!)u_2^2 + (1/4!)u_1^4u_3] + h_6(u_0)(1/5!)u_1^5u_2 + h_7(u_0)(1/7!)u_1^7$$

$$A_8 = h_1(u_0)u_8 + h_2(u_0)[(1/2!)u_4^2 + u_3u_5 + u_2u_6 + u_1u_7]$$

$$+ h_3(u_0)[u_2(1/2!)u_3^2 + (1/2!)u_2^2u_4 + u_1u_3u_4 + u_1u_2u_5 + (1/2!)u_1^2u_6]$$

$$+ h_4(u_0)[(1/4!)u_2^4 + u_1(1/2!)u_2^2u_3 + (1/2!)u_1^2(1/2!)u_3^2$$

$$+ (1/2!)u_1^2u_2u_4 + (1/3!)u_1^3u_5]$$

$$+ h_5(u_0)[(1/2!)u_1^2(1/3!)u_2^3 + (1/3!)u_1^3u_2u_3 + (1/4!)u_1^4u_4]$$

$$+ h_6(u_0)[(1/4!)u_1^4(1/2!)u_2^2 + (1/5!)u_1^5u_3]$$

$$+ h_7(u_0)(1/6!)u_1^6u_2 + h_8(u_0)(1/8!)u_1^8$$

$$A_9 = h_1(u_0)u_9 + h_2(u_0)[u_4u_5 + u_3u_6 + u_2u_7 + u_1u_8]$$

$$+ h_3(u_0)[(1/3!)u_3^3 + u_2u_3u_4 + (1/2!)u_2^2u_5 + u_1(1/2!)u_4^2$$

$$+ u_1u_3u_5 + u_1u_2u_6 + (1/2!)u_1^2u_7]$$

$$+ h_4(u_0)[(1/3!)u_2^3u_3 + u_1u_2(1/2!)u_3^2 + u_1(1/2!)u_2^2u_4$$

$$+ (1/2!)u_1^2u_3u_4 + (1/2!)u_1^2u_2u_5 + (1/3!)u_1^3u_6]$$

$$+ h_5(u_0)[u_1(1/4!)u_2^4 + (1/2!)u_1^2(1/2!)u_2^2u_3$$

$$+ (1/3!)u_1^3(1/2!)u_3^2 + (1/3!)u_1^3u_2u_4 + (1/4!)u_1^4u_5]$$

$$+ h_6(u_0)[(1/3!)u_1^3(1/3!)u_2^3 + (1/4!)u_1^4u_2u_3 + (1/5!)u_1^5u_4]$$

$$+ h_7(u_0)[(1/5!)u_1^5(1/2!)u_2^2 + (1/6!)u_1^6u_3]$$

$$+ h_8(u_0)(1/7!)u_1^7u_2 + h_9(u_0)(1/9!)u_1^9$$

$$A_{10} = h_1(u_0)u_{10} + h_2(u_0)[(1/2!)u_5^2 + u_4u_6 + u_3u_7 + u_2u_8 + u_1u_9]$$

$$+ h_3(u_0)[(1/2!)u_3^2u_4 + u_2(1/2!)u_4^2 + u_2u_3u_5 + (1/2!)u_2^2u_6$$

$$+ u_1u_4u_5 + u_1u_3u_6 + u_1u_2u_7 + (1/2!)u_1^2u_8]$$

$$+ h_4(u_0)[(1/2!)u_2^2(1/2!)u_3^2 + (1/3!)u_2^3u_4 + u_1(1/3!)u_3^3$$

$$+ \ u_1u_2u_3u_4 + u_1(1/2!)u_2^2u_5 + (1/2!)u_1^2(1/2!)u_4^2$$

$$+ \ (1/2!)u_1^2u_3u_5 + (1/2!)u_1^2u_2u_6 + (1/3!)u_1^3u_7]$$

$$+ \ h_5(u_0)[(1/5!)u_2^5 + u_1(1/3!)u_2^3u_3 + (1/2!)u_1^2u_2(1/2!)u_3^2$$

$$+ \ (1/2!)u_1^2(1/2!)u_2^2u_4 + (1/3!)u_1^3u_3u_4$$

$$+ \ (1/3!)u_1^3u_2u_5 + (1/4!)u_1^4u_6]$$

$$+ \ h_6(u_0)[(1/2!)u_1^2(1/4!)u_2^4 + (1/3!)u_1^3(1/2!)u_2^2u_3$$

$$+ \ (1/4!)u_1^4(1/2!)u_3^2 + (1/4!)u_1^4u_2u_4 + (1/5!)u_1^5u_5]$$

$$+ \ h_7(u_0)[(1/4!)u_1^4(1/3!)u_2^3 + (1/5!)u_1^5u_2u_3 + (1/6!)u_1^6u_4]$$

$$+ \ h_8(u_0)[(1/6!)u_1^6(1/2!)u_2^2 + (1/7!)u_1^7u_3]$$

$$+ \ h_9(u_0)(1/8!)u_1^8u_2 + h_{10}(u_0)(1/10!)u_1^{10}$$

Recent studies and analyses have led to development of new computer algorithms to be used for generating the A_n to high orders.

Letting $Nu = u$ yields immediately $A_n = u_n$ for $n = 0,1,2,\ldots$. Since we have pointed out previously that the decomposition method applies to operator equations not necessarily limited to differential operators, let us consider the trivial algebraic equation $x - 8 = 0$ or $x = 8$. Write it as $2x - x - 8 = 0$. Let $Nx = x$ and write $2x - Nx - 8 = 0$. Then $2x = 8 + Nx$ and

$$x = (1/2)8 + (1/2)Nx$$

$$x = 4 + (1/2)[A_0 + A_1 + \ldots]$$

$$x = 4 + (1/2)(4) + (1/2)(2) + \ldots$$

$$= 4 + 2 + 1 + 1/2 + 1/4 + 1/8 + 1/16 + \ldots$$

$$= 7 + \sum_{n=1}^{\infty} (1/2)^n$$

Thus the approximation ϕ_7 to x is given by

$$x = 7 \sum_{n=1}^{4} (1/2)^n = 7 + 1/2 + 1/4 + 1/8 + 1/16 = 7.9375$$

as an approximation to $x = 8$ with ϕ_7 which will evidently improve with more terms.

1.6 Calculation of the A_n Polynomials for Composite Nonlinearities

Some formal definitions will be useful. Let N represent a nonlinear operator and Nx be a nonlinear term in any equation to be solved by decomposition (algebraic, differential, partial differential). A simple term $f(x)$ such as x^2, e^x, or $\sin x$ will be viewed as a zeroth-order composite nonlinearity $\tilde{N}_0 x$ or $N_0 u^0$ where $u^0 = x$, and expanded in the A_n polynomials. We will add a superscript corresponding to the particular nonlinear operator. Thus the A_n will correspond to the N_0 operator, and we have $N_0 u^0 = \sum_{n=0}^{\infty} A_n$.

A first-order composite nonlinearity is written $\tilde{N}_1 x = N_0(N_1 u^1)$ or simply $N_0 N_1 u^1$ where $u^1 = x$ and $u_0 = N_1 u^1$ with $N_0 u^0 = \sum_{n=0}^{\infty} A_n^0$ and $N_1 u^1 = \sum_{n=0}^{\infty} A_n^1$.

For example, the term $e^{-x^2} = \tilde{N}x = N_0 N_1 x$ where $N_0 u^0 = e^{-u} = \sum_{n=0}^{\infty} A_n^0$ and $u^0 = N_1 u^1 = (u^1)^2 = \sum_{n=0}^{\infty} A_n^1$ where $u^1 = x$.

We will emphasize that the superscripts are not exponents; they simply identify the variables and the A_n polynomials for the particular nonlinear operator $N_0, N_1, \dots$.

A second-order composite nonlinearity is written $\tilde{N}_2 x = N_0 N_1 N_2 x$, or $N_0(N_1(N_2 x))$, where

$$N_0 u^0 = \sum_{n=0}^{\infty} A_n^0,$$

$$u^0 = N_1 u^1 = \sum_{n=0}^{\infty} A_n^1,$$

$$u^1 = N_2 u^2 = \sum_{n=0}^{\infty} A_n^2$$

$$u^2 = x.$$

When the decomposition is carried out

$$u^0 = \sum_{n=0}^{\infty} u_n^0$$

$$u^1 = \sum_{n=0}^{\infty} u_n^1$$

$$u^2 = \sum_{n=0}^{\infty} u_n^2.$$

A third-order composite nonlinearity is written

$$\tilde{N}_3x = N_0(N_1(N_2(N_3x))) = N_0N_1N_2N_3x$$

with

$$N_0u^0 = \sum_{n=0}^{\infty} A_n^0$$

$$N_1u^1 = \sum_{n=0}^{\infty} A_n^1$$

$$N_2u^2 = \sum_{n=0}^{\infty} A_n^2$$

$$N_3u^3 = \sum_{n=0}^{\infty} A_n^3$$

and

$$u^3 = x$$

By decomposition

$$u^0 = \sum_{n=0}^{\infty} u_n^0$$

$$u^1 = \sum_{n=0}^{\infty} u_n^1$$

$$u^2 = \sum_{n=0}^{\infty} u_n^2$$

$$u^3 = \sum_{n=0}^{\infty} u_n^3$$

with

$$u^0 = N_1u^1$$

$$u^1 = N_2u^2$$

$$u^2 = N_3u^3$$

$$u^3 = x$$

In general, $N_\nu u^\nu = \sum_{n=0}^{\infty} A_n^{\nu} = u^{\nu-1}$ for $1 \leq \nu \leq m$ with $u^m = x$ and $u^\nu = \sum_{n=0}^{\infty} u_n^{\nu}$.

An mth order composite nonlinearity

$$\tilde{N}_m(x) = N_0(N_1(N_2(\ldots(N_{m-2}(N_{m-1}(N_m(x)))\ldots))) = N_0(u^0) = \sum A_n^0$$

$$N_1(u^1) = \sum A_n^1 = u^0$$

$$N_2(u^2) = \sum A_n^2 = u^1$$

$$N_\nu(u^\nu) = \sum A_n^\nu = u^{\nu-1}$$

$$\vdots$$

$$N_{m-1}(u^{m-1}) = \sum A_n^{m-1} = u^{m-2}$$

$$N_m(u^m) = \sum A_n^m = u^{m-1} \quad \text{with } u^m \equiv x$$

$$\vdots$$

so that the u's are the variables of substitution. Equivalently, $\tilde{N}_m(x) = N_0 \cdot N_1 \cdot N_2 \cdot \ldots N_{m-1} \cdot N_m(x)$, i.e., a composition of operators.

The objective is to determine the A_n polynomials as functions of the x_n's, i.e., $A_n(x_0, x_1, \ldots, x_n) = Nx$.

In the *first-order* case:

$$\begin{aligned} A_n^0 &= A_n^0(x_0, \ldots, x_n) \\ &= A_n^0(u_0^0, \ldots, u_n^0) \\ &= A_n^0(A_0^1, \ldots, A_n^1) \\ &= A_n^0(A_0^1(x_0), \ldots, A_n^1(x_0, \ldots, x_n)). \end{aligned}$$

The *second-order* yields:

$$\overset{0}{A}_n = \overset{0}{A}_n(\overset{0}{u}_0, \ldots, \overset{0}{u}_n)$$
$$= \overset{0}{A}_n(\overset{1}{A}_0, \ldots, \overset{1}{A}_n)$$
$$= \overset{0}{A}_n(\overset{1}{A}_0(\overset{1}{u}_0), \ldots, \overset{1}{A}_n(\overset{1}{u}_0, \ldots, \overset{1}{u}_n))$$
$$= \overset{0}{A}_n(\overset{1}{A}_0(\overset{2}{A}_0), \ldots, \overset{1}{A}_n(\overset{2}{A}_0, \ldots, \overset{2}{A}_n))$$
$$= \overset{0}{A}_n(\overset{1}{A}_0(\overset{2}{A}_0(x_0)), \ldots, \overset{1}{A}_n(\overset{2}{A}_0(x_0), \ldots, \overset{2}{A}_n(x_0, \ldots, x_n)))$$

The *third-order* case yields:

$$\overset{0}{A}_n(\overset{0}{u}_0, \ldots, \overset{0}{u}_n)$$
$$= \overset{0}{A}_n(\overset{1}{A}_0, \ldots, \overset{1}{A}_n)$$
$$= \overset{0}{A}_n(\overset{1}{A}_0(\overset{1}{u}_0), \ldots, \overset{1}{A}_n(\overset{1}{u}_0, \ldots, \overset{1}{u}_n))$$
$$= \overset{0}{A}_n(\overset{1}{A}_0(\overset{2}{A}_0), \ldots, \overset{1}{A}_n(\overset{2}{A}_0, \ldots, \overset{2}{A}_n))$$
$$= \overset{0}{A}_n(\overset{1}{A}_0(\overset{2}{A}_0(\overset{2}{u}_0), \ldots, \overset{1}{A}_n(\overset{2}{A}_0(\overset{2}{u}_0), \ldots, \overset{2}{A}_n(\overset{2}{u}_0, \ldots, \overset{2}{u}_n)))$$
$$= \overset{0}{A}_n(\overset{1}{A}_0(\overset{2}{A}_0(\overset{3}{A}_0)), \ldots, \overset{1}{A}_n(\overset{2}{A}_0(\overset{3}{A}_0), \ldots, \overset{2}{A}_n(\overset{3}{A}_0, \ldots, \overset{3}{A}_n)))$$
$$= \overset{0}{A}_n(\overset{1}{A}_0(\overset{2}{A}_0(\overset{3}{A}_0(x_0))), \ldots, \overset{1}{A}_n(\overset{2}{A}_0(\overset{3}{A}_0(x_0)), \ldots,$$
$$\overset{2}{A}_n(\overset{3}{A}_0(x_0), \ldots \overset{3}{A}_n(_0, \ldots, x_n))))$$

There are a number of ways to handle such composite nonlinearities, but this approach of repeated substitutions appears convenient because it subsumes the limiting (zeroth-order) case of the A_n for Nx and because it appears to be easily programmable.

For an mth order composite nonlinearity, we get

$$\overset{0}{A}_n = \overset{0}{A}_n(\overset{1}{A}_0(\overset{2}{A}_0(\overset{3}{A}_0(\ldots(\overset{\upsilon}{A}_0(\ldots(\overset{m}{A}_0(x_0))\ldots))\ldots))), \ldots,$$
$$\overset{1}{A}_n(\overset{2}{A}_0)(\overset{3}{A}_0(\ldots(\overset{\upsilon}{A}_0(\ldots(\overset{m}{A}_0(x_0))\ldots))\ldots)), \ldots,$$
$$\overset{2}{A}_n(\overset{3}{A}_0(\ldots(\overset{\upsilon}{A}_0(\ldots(\overset{m}{A}_0(x_0))\ldots))\ldots), \ldots,$$

$$\overset{3}{A}_n(\ldots(\overset{\upsilon}{A}_0(\ldots(\overset{m}{A}_0(x_0))\ldots),\ldots,$$

$$\overset{\upsilon}{A}_n(\ldots(\overset{m}{A}_0(x_0),\ldots,\overset{m}{A}_n(x_0,\ldots,x_n))\ldots))\ldots))))$$

Example:

First-order:

$$\tilde{N}_1 x = e^{-\sin(x/2)} = N_0(N_1 x)$$

Let $N_0 u^0 = e^{-\overset{0}{u}} = \sum_{n=0}^{\infty} \overset{0}{A}_n(\overset{0}{u}_0, \overset{0}{u}_1, \ldots, \overset{0}{u}_n)$ and $N_1 u^1 = \sin(u^1/2)$ where $u^1 = x$ and $u^0 = \sum_{n=0}^{\infty} \overset{0}{u}_n = N_1 x = \sin(x/2)$. Calculating the A_n polynomials for the $N_0 u^0$ term (1983; 1986):

$$\overset{0}{A}_0 = e^{-\overset{0}{u}_0}$$

$$\overset{0}{A}_1 = e^{-\overset{0}{u}_0}(-\overset{0}{u}_1)$$

$$\overset{0}{A}_2 = e^{-\overset{0}{u}_0}(-\overset{0}{u}_2 + (1/2)(\overset{0}{u}_1)^2)$$

$$\overset{0}{A}_3 = e^{-\overset{0}{u}_0}(-\overset{0}{u}_3 + \overset{0}{u}_1\overset{0}{u}_2 - (1/6)(\overset{0}{u}_1)^3)$$

$$\vdots$$

(If we omit the identifier superscript, we are dealing with $Nu = e^{-u} = \sum_{n=0}^{\infty} A_n$ where $A_0 = e^{-u_0}$, $A_1 = e^{-u_0}(-u_1)$, etc.) Now calculating the $\overset{1}{A}_n$ for $N_1 x$, we have:

$$\overset{1}{A}_0 = \sin(x_0/2)$$

$$\overset{1}{A}_1 = (x_1/2)\cos(x_0/2)$$

$$\overset{1}{A}_2 = (x_2/2)\cos(x_0/2) - (x_1^2/8)\sin(x_0/2)$$

$$\overset{1}{A}_3 = (x_3/2)\cos(x_0/2) - (x_1 x_2/4)\sin(x_0/2) - (x_1^3/48)\cos(x_0/2)$$

$$\vdots$$

Since $N_0u^0 = \sum_{n=0}^{\infty} A_n^0$ and $u^0 = N_1x = \sum_{n=0}^{\infty} A_n^1$, $u^0 = \sum_{n=0}^{\infty} u_n^0 = \sum_{n=0}^{\infty} A_n^1$,

$$u_0^0 = A_0^1 = \sin(x_0/2)$$

$$u_1^0 = A_1^1 = (x_1/2)\cos(x_0/2)$$

$$u_2^0 = A_2^1 = (x_2/2)\cos(x_0/2) - (x_1^2/8)\sin(x_0/2)$$

$$\vdots$$

Now $N_0u^0 = e^{-u^0} = \sum_{n=0}^{\infty} A_n^0 = A_0^0 + A_1^0 + \ldots = e^{-u_0^0} - u_1^0 e^{-u_0^0} + \ldots$. Thus, now dropping the unnecessary superscript,

$$A_0 = e^{-\sin(x_0/2)}$$

$$A_1 = -(x_1/2)\cos(x_0/2)e^{-\sin(x_0/2)}$$

Any algebraic, differential, or partial differential equation in the author's standard form which contains a nonlinear term $e^{-\sin x/2}$ is now solved by decomposition.

Example:

Second-order

$$\tilde{N}_2x = e^{-\sin^2(x/2)}$$

Let $N_0u^0 = e^{-u^0} = \sum_{n=0}^{\infty} A_n^0$, $N_1u^1 = (u^1)^2 = \sum_{n=0}^{\infty} A_n^1 = u^0 = \sum_{n=0}^{\infty} u_n^0$, and $N_2u^2 = N_2x = \sin(x/2) = \sum_{n=0}^{\infty} A_n^2 = u^1 = \sum_{n=0}^{\infty} u_n^1$. The A_n^0 were specified in the previous example. The A_n^1 are given by:

$$A_0^1 = (u_0^1)^2$$

$$A_1^1 = 2u_0^1u_1^1$$

$$A_2^1 = (u_1^1)^2 + 2u_0^1u_2^1$$

$$A_3^1 = 2u_1^1u_2^1 + 2u_0^1u_3^1$$

$$\vdots$$

and the A_n^2 are:

$$A_0^2 = \sin(x_0/2)$$

$$A_1^2 = (x_1/2)\cos(x_0/2)$$

$$A_2^2 = (x_2/2)\cos(x_0/2) - (x_1^2/8)\sin(x_0/2)$$

$$A_3^2 = (x_3/2)\cos(x_0/2) - (x_1x_2/4)\sin(x_0/2) - (x_1^3/48)\cos(x_0/2)$$

$$\vdots$$

1.7 New Generating Schemes - the Accelerated Polynomials

We have previously pointed out that the A_n polynomials are not unique. Certain terms can be moved forward or back, i.e., to the component u_n for higher or lower n. This affects the convergence rate since n-term approximations are sought. For example, if $Nu = f(u) = u^2$, we have

$$A_0 = u_0^2$$

$$A_1 = 2u_0u_1$$

$$A_2 = u_1^2 + 2u_0u_2$$

$$A_3 = 2u_1u_2 + 2u_0u_3$$

If the u_1^2 term were moved from A_2 to A_1, the solution $u = \sum_{n=0}^{\infty} u_n$ would be unaffected, but an approximate solution which only went as far as the u_2 term would be different - actually closer to the final result. The scheme for generating the A_n was chosen so that the sum of the subscripts of each term in A_n would be n. This was done for consistency between mathematical definition and an earlier heuristic treatment.

In the new scheme we will now propose, certain terms of the A_n are moved up which accelerates the convergence; consequently, they are called the *accelerated polynomials* and denoted by $\hat{A}_n$ to avoid confusion with the A_n. The principal reason for developing this form was not the improvement in an already rapid convergence, but the avoidance of the parametrization which causes the decomposition series to appear to be a perturbation procedure which is an incorrect conclusion. Eliminating it yields a more physical derivation. Despite these advantages, it still may sometimes be desirable to use the A_n rather than the $\hat{A}_n$. It is sometimes easier to see the sum to which the approximation series converges when using the A_n. Also in some problems involving differential equations, integrability difficulties arise with the $\hat{A}_n$. Thus, this is still an area of study, and both forms may be useful in applications as well as being mathematically interesting. Now we derive the $\hat{A}_n$.

In the parametrized decomposition, evaluation of $u|_{\lambda=0}$ results in u_0, and derivatives with respect to λ will be replaced by derivatives with respect to u_0. Noting again the earlier formula

$$A_n = 1/n! \ (d^n/ d\lambda^n) f(u(\lambda))|_{\lambda=0}$$

and the fact that the nth term of the series for e^x is $x^n/n!$, we define the operator exponential ξ_n by

$$\xi_n = e^{u \, d/du_0} - 1 = \sum_{\upsilon=1}^{\infty} u^{\upsilon}/\upsilon!)(d^{\upsilon}/du_0^{\upsilon}) \qquad (1.7.1)$$

Exercise:

In *Nonlinear Stochastic Operator Equations* (1986), the "forward difference operator," used in finite difference methods, was defined by

$$\eta = \sum_{n=1}^{\infty}(\tau^n/n!)d^n/dt^n$$

Replacing t by u_0 and τ by u_n, and n by υ, we have an interesting analogy. In (1.6.1) the ξ_n is an operator. In terms of the ξ, we can define an operator A by

$$A_0 = 1 \text{ (identity operator)}$$
$$A_{n \geq 1} = \xi_n \sum_{\upsilon=0}^{n-1} A_\upsilon \qquad (1.7.2)$$

Thus,

$$A_0 = 1$$

$$A_1 = \xi_1 A_0 = \xi_1 = e^{u_1 d/du_0} - 1$$

$$A_2 = \xi_2(A_0 + A_1) = \xi_2(1 + \xi_1) = \xi_2 + \xi_2\xi_1$$

$$A_3 = \xi_3(A_0 + A_1 + A_2) = \xi_3(1 + \xi_1 + \xi_2 (1 + \xi_1))$$

$$A_4 = \xi_4(A_0 + A_1 + A_2 + A_3)$$

$$= \xi_4(1 + \xi_1 + \xi_2(1 + \xi_1) + \xi_3(1 + \xi_1 + \xi_2(1 + \xi_1))$$

$$\vdots$$

We note that $\hat{A}_0 \equiv A_0 = f(u_0)$ so that both forms of the polynomials are identical at $n = 0$.

If we represent d/du_0 simply by D, we can write:

$$\hat{A}_0 = f(u_0)$$

$$\hat{A}_1 = (e^{u_1 D} - 1)f(u_0)$$

$$\hat{A}_2 = (e^{u_2 D} - 1)f(u_0) + (e^{u_2 D} - 1)(e^{u_1 D} - 1)f(u_0)$$

$$\hat{A}_3 = (e^{u_3 D} - 1)f(u_0) + (e^{u_3 D} - 1)(e^{u_2 D} - 1)f(u_0)$$

$$+ (e^{u_3 D} - 1)(e^{u_1 D} - 1)f(u_0)$$

$$+ (e^{u_3 D} - 1)(e^{u_2 D} - 1)(e^{u_1 D} - 1)f(u_0)$$

$$\hat{A}_4 = \{(e^{u_4 D} - 1) + (e^{u_4 D} - 1)(e^{u_3 D} - 1) \qquad (1.7.3)$$

$$+ (e^{u_4 D} - 1)(e^{u_2 D} - 1) + (e^{u_4 D} - 1)(e^{u_1 D} - 1)$$

$$+ (e^{u_4 D} - 1)(e^{u_3 D} - 1)(e^{u_2 D} - 1)$$

$$+ (e^{u_4 D} - 1)(e^{u_3 D} - 1)(e^{u_1 D} - 1)$$

$$+ (e^{u_4 D} - 1)(e^{u_2 D} - 1)(e^{u_1 D} - 1)$$

$$+ (e^{u_4 D} - 1)(e^{u_3 D} - 1)(e^{u_2 D} - 1)(e^{u_1 D} - 1)\}f(u_0)$$

etc. The accelerated polynomials $\hat{A}_n$ can now be defined as:

$$\hat{A}_n = \mathcal{A}_n f(u_0) \tag{1.7.4}$$

in terms of (1.6.1) and (1.6.2). This is in many ways a more satisfactory definition.

Example:

To solve the algebraic equation $x = k + e^{-x}$ with $k = 2$, we write the equation in the form $x = \sum_{n=0}^{\infty} x_n = k + \sum_{n=0}^{\infty} \hat{A}_n$ where the $\hat{A}_n$ are evaluated for $Nx = e^{-x}$ and $k = 2$. Now

$$\hat{A}_0 = f(x_0) = e^{-x_0}$$

$$\hat{A}_1 = \xi_1 e^{-x_0} = (e^{x_1 d/dx_0} - 1) e^{-x_0} = (e^{-x_1} - 1) e^{-x_0}$$

$$\hat{A}_2 = [\xi_2(1 + \xi_1)] e^{-x_0} = (e^{-x_2} - 1) [1 + (e^{-x_1} - 1)]e^{-x_0}$$

$$\hat{A}_3 = \{(e^{-x_3} - 1)[1 + (e^{-x_1} - 1) + (e^{-x^2} - 1)(1 + (e^{-x_1} - 1)]\}e^{-x_0}$$

.
.
.

Now since $x_0 = 2.0$ and $x_{n \geq 1} = \hat{A}_{n-1}$, we have

$$x_0 = 2$$

$$x_1 = .1353352832$$

$$x_2 = -.0171303316$$

$x_3 = .002042333$

Since an n-term approximation to x is given by $\phi_n = \sum_{\upsilon=0}^{n-1} x_\upsilon$,

$\phi_1 = 2.00 \ldots$

$\phi_2 = 2.135335283$

$\phi_3 = 2.118204952$

$\phi_4 = 2.120247285$

The correct solution is 2.120028239. Defining ψ_n as the % error in n terms

$\psi_1 = -5.66\%$

$\psi_2 = +0.722\%$

$\psi_3 = -0.08600\%$

$\psi_4 = +0.01033\%$

Using the A_n polynomials instead of the $\hat{A}_n$, the results are:

$\psi_1 = -5.66\%$

$\psi_2 = 0.722\%$

$\psi_3 = -0.142\%$

$\psi_4 = 0.033\%$

We see that use of the (accelerated polynomials) $\hat{A}_n$ results in faster convergence.

We can define $\psi_m = [(\phi_m - x)/x](100)$ and $\hat{\psi}_m = [(\hat{\phi}_m - x)/x](100)$ and the acceleration ratio $\rho_m = [(\psi_m - \hat{\psi}_m)/\psi_m](100)$ and note that acceleration is 68.7% by the fourth term. Thus the new definition appears to be a convenient one. It results in still faster convergence. It is also more physical; differentiations are with respect to the physical term u_0 instead of the parameter λ.

Thus, just as a Fourier series expresses a function in terms of trigonometric functions, we have expressed a function in terms of a rapidly converging set of polynomials. This is one essential step in the solution by decomposition of general nonlinear equations. The other is to write the equation in the form $Lu + Nu = g$. If L is invertible, solve for Lu thus:

$$Lu = g - Nu$$

Now operating with L^{-1}, we write

$$u = \Phi + L^{-1}g - L^{-1}Nu = \sum_{n=0}^{\infty} u_n$$

where $L\Phi = 0$, identify $u_0 = \Phi + L^{-1}g$ and replace Nu by $\sum_{n=0}^{\infty} A_n$ or, equivalently $\sum_{n=0}^{\infty} \hat{A}_n$ and identify terms. Thus u_0 is known and

$$u_1 = -L^{-1}A_0$$

$$u_2 = -L^{-1}A_1$$

$$\vdots$$

If the linear operator is not invertible, or if it is not easily invertible, we will write

$$Lu + Ru + Nu = g$$

where L is invertible and R is the remainder of the linear operator, and then, proceed as before. Now

$$u_1 = -L^{-1}Ru_0 - L^{-1}A_0$$

$$u_2 = -L^{-1}Ru_1 - L^{-1}A_1$$

$$\vdots$$

The procedure will solve algebraic equations, differential equations, integro-differential equations, and with some further considerations, systems of differential equations, partial differential equations, and systems of partial differential equations (1983; 1986).

Example:

$$f(u) = u^2$$

$$\hat{A}_0 = f(u_0) = u_0^2$$

$$\hat{A}_1 = (e^{u_1 D} - 1)f(u_0) = (u_1\, d/du_0 + (u_1^2/2)d^2/du_0^2 + \ldots)u_0^2$$

$$= u_1^2 + 2u_1u_0$$

$$\hat{A}_2 = (e^{u_2 D} - 1)u_0^2 + (e^{u_2 D} - 1)\,(e^{u_1 D} - 1)u_0^2$$

$$= u_2^2 + 2u_2(u_0 + u_1)$$

$$\vdots$$

i.e., $\hat{A}_{n\geq 1} = u_n^2 + 2u_n \sum_{\upsilon=0}^{n-1} u_\upsilon$.

Suppose we consider the differential equation which incorporates $f(u) = u^2$ such as $du/dt = u^2$ with $u(0) = 1$. Using decomposition,

$$u = \sum_{n=0}^{\infty} u_n = u(0) + L^{-1} \sum_{n=0}^{\infty} \hat{A}_n$$

Thus $u_0 = 1$ and $u_{n\geq 1} = L^{-1}\{u_{n-1}^2 + 2\, u_{n-1} \sum_{\upsilon=0}^{n-2} u_\upsilon\}$ or,

$$u_1 = L^{-1}(1) = t$$

$$u_2 = L^{-1}\,\hat{A}_1 = L^{-1}(t^2 + 2t) = (t^3/3) + t^2$$

$$u_2 = L^{-1}\,\hat{A}_1 = L^{-1}(t^2 + 2t) = (t^3/3) + t^2$$

$$u_3 = L^{-1}\,\hat{A}_2 = L^{-1}[(t^6/9) + (2t^5/3) + (t^4) + (2t^4/3) + (2t^3)]$$

$$= (t^7/63) + (t^6/9) + (t^5/3) + (2t^4/3) + (2t^3/3)$$

At $t = 1/2$ we have

$u_0 = 1$

$u_1 = 0.500$

$u_2 = 0.2916667$

$u_3 = 0.1372768$

and the n-term approximation

$\phi_1 = 1.00$

$\phi_2 = 1.50$

$\phi_3 = 1.7916667$

$\phi_4 = 1.9289435$

The exact answer is $u = 1/(1 - t)$, hence $u(1/2) = 2$ and we see $\phi_n \to 2$. Had we used A_n instead of $\hat{A}_n$, the result converges more slowly. Thus

$\phi_1 = 1.00$

$\phi_2 = 1.50$

$\phi_3 = 1.75$

$\phi_4 = 1.875$

Returning to the result for $u(t)$ and using the accelerated polynomials $\hat{A}_n$, we have

$u_0 = 1$

$u_1 = t$

$u_2 = (t^3/3) + t^2$

$$u_3 = (2t^3/3) + (2t^4/3) + (t^5/3) + (t^6/9) + (t^7/63)$$

where $\sum_{n=0}^{\infty} u_n$ is the solution. With the A_n, we have

$$u_0 = 1$$

$$u_1 = t$$

$$u_2 = t^2$$

$$u_3 = t^3$$

which clearly is a geometric series. Thus $u(t) = \sum_{n=0}^{\infty} t^n = 1/(1 - t)$ which is less obvious in the solution with $\hat{A}_n$ although we might notice that the first terms of u_2 and u_3 add to give t^3, etc.

Example:

$$f(u) = e^u$$

Now $f(u_0) = e^{u_0}$. Then $\xi_n e^{u_0} = (e^{u_n d/du_0} - 1)e^{u_0} = (e^{u_n} - 1)e^{u_0}$ i.e., $\hat{A}_n = (e^{u_n} - 1)e^{u_0}$ for $n \geq 1$ and $\hat{A}_0 = e^{u_0}$.

Example:

$$f(u) = e^{-u}$$

Since $(d^{\upsilon}/du^{\upsilon})e^{-u_0} = (-1)^{\upsilon}\, e^{-u_0}$, then

$$\xi_m\, e^{-u_0} = \sum_{\upsilon=1}^{\infty} u_m^{\upsilon}/\upsilon!\; d^{\upsilon}/du_0^{\upsilon}\, e^{-u_0} = \sum_{\upsilon=1}^{\infty} u_m^{\upsilon}/\upsilon!\, (-1)^{\upsilon}\, e^{-u_0}$$

$$= e^{-u_0}\, [\sum_{\upsilon=1}^{\infty} (u_m^{\upsilon}/\upsilon!)\, (-1)^{\upsilon}] = e^{-u_0}(e^{-u_m} - 1).$$

Example:

$$f(u) = e^{-\alpha u}$$

Now $f(u_0) = e^{-\alpha u_0}$ and

$$\sum_{\upsilon=1}^{\infty} u_m^\upsilon/\upsilon!\, d^\upsilon/du_0^\upsilon\, e^{-\alpha u_0} = \sum_{\upsilon=1}^{\infty} u_m^\upsilon/\upsilon!\, (-1)^\upsilon \alpha^\upsilon e^{-\alpha u_0}$$

$$= e^{-\alpha u_0} [\sum_{\upsilon=1}^{\infty} u_m^\upsilon/\upsilon!\, (-1)^\upsilon \alpha^\upsilon] = e^{-\alpha u_0}[e^{-\alpha u_0} - 1]$$

Listing the accelerated form $\hat{A}_n$ we have

$$\hat{A}_0 = f(u_0)$$

$$\hat{A}_1 = (e^{u_1 d/du_0} - 1)f(u_0)$$

$$A_2 = (e^{u_2 d/du_0} - 1)f(u_0)$$

$$+ (e^{u_2 d/du_0} - 1)(e^{u_1 d/du_0} - 1)f(u_0)$$

$$\hat{A}_3 = (e^{u_3 d/du_0} - 1)f(u_0)$$

$$+ (e^{u_3 d/du_0} - 1)(e^{u_2 d/du_0} - 1)f(u_0)$$

$$+ (e^{u_3 d/du_0} - 1)(e^{u_1 d/du_0} - 1)f(u_0)$$

$$+ (e^{u_3 d/du_0} - 1)(e^{u_2 d/du_0} - 1)(e^{u_1 d/du_0} - 1)f(u_0)$$

$$\vdots$$

or

$$\hat{A}_0 = f(u_0)$$

$$\hat{A}_1 = \xi_1 f(u_0)$$

$$\hat{A}_2 = \xi_2(1 + \xi_1) f(u_0) = (\xi_2 + \xi_2\xi_1)f(u_0)$$

$$\hat{A}_3 = \xi_3(1 + \xi_1 + \xi_2(1 + \xi_1))f(u_0)$$

$$= (\xi_3 + \xi_3\xi_1 + \xi_3\xi_2 + \xi_3\xi_2\xi_1)f(u_0)$$

$$A_4 = \xi_4(1 + \xi_1 + \xi_2(1 + \xi_1) + \xi_3(1 + \xi_1 + \xi_2(1 + \xi_1)))f(u_0)$$

$$= (\xi_4 + \xi_4\xi_1 + \xi_4\xi_2 + \xi_4\xi_3 + \xi_4\xi_2\xi_1$$

$$+ \xi_4\xi_3\xi_1 + \xi_4\xi_3\xi_2 + \xi_4\xi_3\xi_2\xi_1)f(u_0$$

.
.
.

Or it may be convenient to use $\hat{A}_n = A_n f(u_0)$ where $A_0 = 1$, the identity operator, and $A_{n\geq 1} = \xi_n \sum_{\upsilon=0}^{n-1} A_\upsilon$ as in (1.6.2) where $A_1 = \xi_1$, $A_2 = \xi_2(1+\xi_1)$, $A_3 = \xi_3(1+\xi_1+\xi_2(1+\xi_1))$, etc. as previously given.

A comparison of the A_n and $\hat{A}_n$ for $f(u) = u^2$ yields

$$\hat{A}_0 = u_0^2$$

$$\hat{A}_1 = u_1^2 + 2u_0u_1$$

$$\hat{A}_2 = u_2^2 + 2u_0u_2 + 2u_1u_2$$

$$\hat{A}_3 = u_3^2 + 2u_0u_3 + 2u_1u_3 + 2u_2u_3$$

$$\hat{A}_4 = u_4^2 + 2u_0u_4 + 2u_1u_4 + 2u_2u_4 + 2u_3u_4$$

.
.
.

$$A_m = u_m^2 + 2u_m \sum_{\upsilon=0}^{m-1} u_\upsilon$$

and

$$A_0 = u_0$$

$$A_1 = 2u_0u_1$$

$$A_2 = u_1^2 + 2u_0u_2$$

$$A_3 = 2u_0u_3 + 2u_1u_2$$

$$A_4 = u_2^2 + 2u_0u_4 + 2u_1u_3$$

We see that the terms involving u_n arrive later in the A_n, i.e., their appearance in the $\hat{A}_n$ is accelerated over that of the A_n. In the A_2, u_1^2 arrives late as compared with $\hat{A}_2$. In A_3, $2u_1u_2$ arrives late, etc. Now consider the $\hat{A}_n$ for $f(u) = u^3$:

$$\hat{A}_0 = u_0^3$$

$$\hat{A}_1 = u_1^3 + 3u_0^2u_1 + 3u_1^2u_0$$

$$\hat{A}_2 = u_2^3 + 3u_0^2u_2 + 3u_1^2u_2 + 3u_2^2u_0 + 3u_2^2u_1 + 6u_0u_1u_2$$

$$\hat{A}_3 = u_3^3 + 3u_0^2u_3 + 3u_1^2u_3 + 3u_2^2u_3 + 3u_3^2u_0 + 3u_3^2u_1$$
$$+ 3u_3^2u_2 + 6u_0u_1u_3 + 6u_0u_2u_3 + 6u_1u_2u_3$$

$$\hat{A}_4 = u_4^3 + 3u_0^2u_4 + 3u_1^2u_4 + 3u_2^2u_4 + 3u_3^2u_4 + 3u_4^2u_0 + 3u_4^2u_1$$
$$+ 3u_4^2u_2 + 3u_4^2u_3 + 6u_0u_1u_4 + 6u_0u_2u_4 + 6u_0u_3u_4$$
$$+ 6u_1u_2u_4 + 6u_1u_3u_4 + 6u_2u_3u_4$$

.
.
.

$$\hat{A}_m = u_m^3 + 3u_m \sum_{\upsilon=0}^{m-1} u_\upsilon^2 + 3u_m^2 \sum_{\upsilon=0}^{m-1} u_\upsilon$$
$$+ 6u_m \sum_{\mu=\upsilon+1}^{m-1} \sum_{\upsilon=0}^{m-2} u_\mu u_\upsilon$$

Now consider $f(u) = u^4$. We have $\hat{A}_0 = u^4$ and for $m \geq 1$ we define

$$\hat{A}_{m\geq 1} = u_m^4 + 4u_m \sum_{\upsilon=0}^{m-1} u_\upsilon^3 + 4u_m^3 \sum_{\upsilon=0}^{m-1} u_\upsilon$$
$$+ 6u_m^2 \sum_{\upsilon=0}^{m-1} u_\upsilon^2 + 12u_m \sum_{\mu=\upsilon+1}^{m-1} u_\mu^2 \sum_{\upsilon=0}^{m-2} u_\upsilon$$
$$+ 12u_m \sum_{\mu=\upsilon+1}^{m-1} u_\mu \sum_{\upsilon=0}^{m-2} u_\upsilon^2$$
$$+ 12u_m^2 \sum_{\mu=\upsilon+1}^{m-1} u_\mu \sum_{\upsilon=0}^{m-2} u_\upsilon$$
$$+ 24u_m \sum_{\mu=\upsilon+1}^{m-1} u_\mu \sum_{\mu=\Upsilon+1}^{m-2} u_\upsilon \sum_{\Upsilon=0}^{m-3} u_\Upsilon$$

It is interesting to consider briefly the Maclaurin expansion for $f(x)$. We have

$$f(x) = f(0) + xf^{(1)}(0) + (x^2/2!)f^{(2)}(0) + \cdots$$

$$= \sum_{n=0}^{\infty} (x^n/n!) f^{(n)}(0)$$

$$= \sum_{n=0}^{\infty} (x^n/n!)(d^n/dx^n)\, f(x)\big|_{x=0}$$

Thus we can define the Maclaurin operator M as

$$M = \sum_{n=0}^{\infty} (x^n/n!) d^n/dx^n \equiv e^{x d/dx}$$

in terms of the operator exponential. We can also write

$$M = 1 + \sum_{n=1}^{\infty} (x^n/n!)(d^n/dx^n)$$

or

$$M = 1 + (\, e^{x d/dx} - 1)$$

for comparison with our operator A from which the $\hat{A}_n$ (accelerated polynomials) are obtained, i.e., $A\, f(u)\big|_{u=u_0}$ in our series and $Mf(x)\big|_{x=0}$ in the Maclaurin series. The operator A is given by $\sum_{n=0}^{\infty} A_n$ where the A_n are given by:

$$A_0 = 1$$

$$A_1 = \xi_1$$

$$A_2 = \xi_2(1 + \xi_1)$$

$$A_3 = \xi_3(1 + \xi_1 + \xi_2(1 + \xi_1))$$

$$\vdots$$

or

$$A = 1 + (e^{u_1 d/du_0} - 1) + (e^{u_2 d/du_0} - 1)(1 + (e^{u_1 d/du_0} - 1)) + \dots$$

$$A = 1 + (e^{u_1 d/du_0} - 1) + (e^{u_2 d/du_0} - 1)$$

$$+ (e^{u_2 d/du_0} - 1)\,(e^{u_1 d/du_0} - 1)$$

$$+ (e^{u_3 d/du_0} - 1) + (e^{u_3 d/du_0} - 1)\,(e^{u_2 d/du_0} - 1)$$

$$+ (e^{u_3 d/du_0} - 1)(e^{u_1 d/du_0} - 1)$$

$$+ (e^{u_3 d/du_0} - 1)(e^{u_2 d/du_0} - 1)(e^{u_1 d/du_0} - 1)$$

$$+ \cdots$$

which clearly is much more complex than the Maclaurin series.

$$\begin{aligned} A &= 1 + \xi_1 + \xi_2 + \xi_2\xi_1 + \xi_3 + \xi_3\xi_2 \\ &\quad + \xi_3\xi_1 + \xi_3\xi_2\xi_1 + \xi_4 + \xi_4\xi_1 + \xi_4\xi_2 \\ &\quad + \xi_4\xi_3 + \xi_4\xi_2\xi_1 + \xi_4\xi_3\xi_1 \\ &\quad + \xi_4\xi_3\xi_2 + \xi_4\xi_3\xi_2\xi_1 + \cdots \end{aligned}$$

Note that the operands are quite different also

$M\, f(x)|_{x=0}$ for Maclaurin series

$A\, f(u)|_{u=u_0}$ for Adomian series

The $M\, f(x)|_{x=0}$ and u are functions of x while $A\, f(u)|_{u=u_0}$ is a function of $u_0, u_1, \ldots$. In solving a differential equation such as $du/dx + f(u) = 0$, our decomposition parameter λ is not a parameter appearing in the differential equation, but the differentiations for computation of the A_n are done with respect to λ. This is not the case with the Maclaurin series (where x, u, f are intrinsic parameters).

Note if we write $\hat{A}_0 = f(u_0) = h(u_0)$ as in (1.1.2) and write

$$h_m = d^m/du_0^m\, f(u_0) = h_m(u_0)$$

and

$$\begin{aligned} \hat{A}_1 &= e^{(u_1 d/du_0} - 1)f(u_0) \\ &= [(u_1(d/du_0) + u_1^2/2!)(d^2/du_0^2) + \cdots)]f(u_0) \\ &= u_1\, h_1(u_0) + (u_1^2/2!)h_2(u_0) + (1/3!)u_1^3\, h_3(u_0) + \cdots \end{aligned}$$

$$\hat{A}_2 = \{(e^{u_2 d/du_0} - 1) + (e^{u_2 d/du_0} - 1)(e^{u_1 d/du_0} - 1)\}f(u_0)$$

$$= u_2\, h_1(u_0) + (u_2^2/2!)h_2(u_0) + (1/3!)\, u_2^3\, h_3(u_0)$$

$$+ \ldots + u_1u_2h_2(u_0) + (1/2!)(u_1u_2^2\, h_3(u_0)$$

$$+ \ldots + 1/2!\; u_2^2u_1\; h_3(u_0) + (1/2!)\; u_1^2\; (1/2!)\; u_2^2\; h_4(u_0)$$

$$+ (1/3!)\; u_1^3u_2\; h_4(u_0) + (1/3!)\; u_2^3u_1\; h_4(u_0)$$

$$+ (1/4!)\; u_1^4\; u_2\; h_5(u_0) + (1/2!)\; u_1^2\; (1/3!)\; u_2^3\; h_5(u_0)$$

$$+ u_1\; (1/4!)\; u_2^4\; h_5(u_0)$$

Consider the term

$$(e^{u_2 d/du_0} - 1)(e^{u_1 d/du_0} - 1)f(u_0)$$

$$= (\textstyle\sum_{m=\upsilon}^{\infty} u^m/m!\; (d/du_0)^m - 1)(\; \sum_{m=\upsilon}^{\infty}(u_1^m/m!)(d/du_0)^m - 1)f(u_0)$$

$$= u_1u_2\; h_2(u_0) + [(u_1^2/2!)\; u_2 + (u_1u_2^2/2!)]h_3(u_0)$$

$$+ [(u_1^3/3!)u_2 + (1/2!)\; u_1^2\; (1/2!)\; u_2^2 + (u_2^3/3!)\; u_1]h_4(u_0)$$

$$+ [(1/4!)\; u_1^4u_2 + (1/3!)\; u_1^3\; (1/2!)\; u_2^2$$

$$+ (1/2!)\; u_1^2\; (1/3!)\; u_2^3 + u_1\; (1/4!)\; u_2^4]h_5(u_0) + \cdots$$

Thus,

$$\hat{A}_2 = (e^{u_2 d/du_0} - 1)f(u_0)$$

$$+ (e^{u_2 d/du_0} - 1)(e^{u_1 d/du_0} - 1)f(u_0)$$

so we see that the $\hat{A}_n$, i.e., the accelerated polynomials, are evaluated for comparison with the A_n previously derived:

$$\hat{A}_1 = u_1h_1 + (u_1^2/2!)\; h_2 + (u_1^3/3!)\; h_3$$

while

$$A_1 = u_1 h_1$$

Also, $A_2 = u_2 h_1 + (1/2!)\, u_1^2\, h_2$ can be compared with $\hat{A}_2$ above. It is clear that more and more terms are moved forward in the $\hat{A}_n$ and so an n-term approximation using $\hat{A}_n$ would contain more of the series. With $f(u) = u^2$

$$\hat{A}_0 = f(u_0) = u_0^2$$

$$\hat{A}_1 = (e^{u_1 d/du_0} - 1)f(u_0) = u_1(d/du_0)\, f(u_0) + (u_1^2/2!)(d^2/du_0^2)f(u_0)$$

where the series terminates since $f(u_0) = u^2$. Hence

$$\hat{A}_1 = 2u_0 u_1 + u_1^2$$

$$\hat{A}_2 = (e^{u_2 d/du_0} - 1)f(u_0) + (e^{u_2 d/du_0} - 1)(e^{u_1 d/du_0} - 1)f(u_0)$$

$$= u_2\, d/du_0\, f(u_0) + u_2^2/2!\, d^2/du_0^2\, f(u_0)$$

$$+ [u_2\, (d/du_0) + (u_2^2/2!)\, (d^2/du_0^2)][u_1(d/du_0)$$

$$+ (u_1^2/2!)(d^2/du_0^2)]f(u_0)$$

$$= 2u_0 u_2 + u_2^2 + 2u_1 u_2$$

etc. Again, these involve more terms than the A_n. For reference

$$\hat{A}_3 = u_3\, h_1 + u_1 u_2\, h_2 + (1/3!)\, u_1^3 h_3$$

$$\hat{A}_4 = u_4\, h_1 + ((1/2!)\, u_2^2 + u_1 u_3)h_2$$

$$+ (1/2!)\, (u_1^2 u_2)h_3 + (1/4!)\, u_1^4\, h_4$$

$$\hat{A}_5 = u_5\, h_1 + (u_2 u_3 + u_1 u_4)h_2 + (u_1\, (1/2!)\, u_2^2 + (1/2!)u_1^2 u_3)h_3$$

$$+ (1/3!)\, u_1^3 u_2\, h_4 + (1/5!)\, u_1^5\, h_5$$

$$\hat{A}_6 = u_6 h_1 + ((1/2!)\, u_3^2 + u_2u_4 + u_1u_5)h_2$$
$$+ (1/3!)(u_2^3 + u_1u_2u_3 + (1/2!)\, u_1^2u_4)h_3$$
$$+((1/2!)u_1^2(1/2!)\, u_2^2 + (1/3!)\, u_1^3u_3)h_4$$
$$+ (1/4!)\, u_1^4u_2h_5 + (1/6!)\, u_1^6\, h_6$$

etc.

1.8 Convergence of the A_n Polynomials

Given a function $f(u)$ with the necessary differentiability, we write

$$f(u) = \sum_{n=0}^{\infty} \hat{A}_n$$

We have written $\hat{A}_n$ to distinguish these from the A_n. Given $f(u)$, we have $f(u_0)$, and we require

$$\hat{A}_n = \mathcal{A}_n f(u_0)$$

where

$$\hat{A}_n = \xi_n \sum_{\upsilon=0}^{n-1} \mathcal{A}_\upsilon$$

for $n \geq 1$, $\mathcal{A}_0 \equiv 1$, and

$$\xi_n = e^{u_n d/du_0} - 1 = \sum_{\upsilon=1}^{\infty} (u^\upsilon/\upsilon!) d^\upsilon/du^\upsilon \equiv \sum_{\upsilon=1}^{\infty} c_n^\upsilon$$

i.e. $c_n = u^\upsilon/\upsilon!\; d^\upsilon/du^\upsilon$. Thus $\mathcal{A}_n$ and ξ_n are operators. We have now

$$\mathcal{A}_0 = 1$$

$$\mathcal{A}_1 = \xi_1 \mathcal{A}_0 = \xi_1 = e^{u_1 d/du_0} - 1$$

$$\mathcal{A}_2 = \xi_2(\mathcal{A}_0 + \mathcal{A}_1) = \xi_2(1 + \xi_1) = \xi_2 + \xi_2\xi_1$$

$$\mathcal{A}_3 = \xi_3(\mathcal{A}_0 + \mathcal{A}_1 + \mathcal{A}_2) = \xi_3(1 + \xi_1 + \xi_2(1 + \xi_1))$$

$$A_4 = \xi_4(A_0 + A_1 + A_2 + A_3)$$

$$= \xi_4(1 + \xi_1 + \xi_2(1 + \xi_1) + \xi_3(1 + \xi_1 + \xi_2(1 + \xi_1)))$$

$$\vdots$$

We now have

$$\sum_{n=0}^{\infty} A_n f(u_0) = [1 + \sum_{i=1}^{\infty} \xi_i + \sum_{\substack{i,j=1 \\ i\neq j}}^{\infty} \xi_i\xi_j + \sum_{\substack{i,j,k=1 \\ i\neq j\neq k}}^{\infty} \xi_i \xi_j \xi_k + \ldots] f(u_0)$$

Equivalently,

$$\sum_{n=0}^{\infty} A_n f(u_0) = f(u_0) + \sum_{i=1}^{\infty}(e^{u_i d/du_0} - 1)f(u_0) + \sum_{\substack{i,j=1 \\ i\neq j}}^{\infty} (e^{u_i d/du_0} - 1)(e^{u_j d/du_0} - 1)f(u_0) + \ldots$$

Thus

$$\sum_{n=0}^{\infty} \hat{A}_n = f(u_0) + \sum_{i=1}^{\infty}\{u_i df(u_0)/du_0 + (u_i^2/2!)\, d^2f(u_0)/du_0^2 + \cdots\} + \ldots$$

$$= f(u_0) + \{u_1 + u_2 + \cdots\}\, df(u_0)/du_0 + \{(u_1^2/2!) + u_1u_2 + \ldots\}\, d^2 f(u_0)/du_0^2 + \cdots$$

$$= f(u_0) + (u - u_0)\, df(u_0)/du_0 + (1/2!)(u - u_0)^2\, d^2f(u_0)/du_0^2 + \cdots$$

$$= \sum_{n=0}^{\infty} [(u - u_0)^n/n!] d^nf(u_0)/du_0^n$$

which is the Taylor expansion of $f(u)$ about $u_0 = u_0(x)$ which is the initial term (a function, not merely a constant). Thus we can state the theorem:

$$f(u) = \sum_{n=0}^{\infty} \hat{A}_n$$

in the neighborhood of $u_0(x)$ with the $\hat{A}_n$ defined as above.

We see that if Nu is a polynomial nonlinearity, not only do the higher derivatives vanish, but we have $n!$ in the denominator of the ξ_i terms and products of $n!$ in the product terms $\xi_i\xi_j$, $\xi_i\xi_j\xi_k$, etc. Thus ξ_i terms approach zero as $1/n!$ and the product terms approach zero as the inverse of products of factorials. For an equation $Lu + Nu = g$ with Nu expressed in the A_n or $\hat{A}_n$, we get computable series which depend only on n components rather than an infinite number. The A_n are not the same as the $\hat{A}_n$, but $\sum_{n=0}^{\infty} A_n = \sum_{n=0}^{\infty} \hat{A}_n$.

Example:

Consider the linear limit $f(u) = u$. We have $f(u_0) = u_0$

$$f(u) = \sum_{n=0}^{\infty} \hat{A}_n = \sum_{n=0}^{\infty} \mathcal{A}_n f(u_0) = \sum_{n=0}^{\infty} \mathcal{A}_n u_0$$

$$= \mathcal{A}_0 u_0 + \mathcal{A}_1 u_0 + \ldots$$

where

$$\mathcal{A}_0 u_0 = u_0$$

$$\mathcal{A}_1 u_0 = (e^{u_1 d/du_0} - 1)u_0$$

$$= (u_1 d/du_0 + \cdots)u_0 = u_1$$

$$\mathcal{A}_2 u_0 = \xi_2 u_0 + \xi_2\xi_1 u_0$$

$$= (u_2 d/du_0 + \cdots)u_0$$

$$+ (u_2 d/du_0 + \cdots)(u_1 d/du_0 + \cdots)u_0 = u_2$$

$$\vdots$$

Thus $f(u) = u_0 + u_1 + u_2 + \cdots = \sum_{n=0}^{\infty} u_n = u$ and the $\hat{A}_n$ reduce to u_n in the linear case $f(u) = u$.

For the case $f(u) = u^2$, we obtain

$$A_0 f(u_0) = u_0^2$$

$$A_1 f(u_0) = (u_1 d/du_0 + \cdots)u_0 = 2u_0u_1 + u_1^2$$

etc.

We note that $\hat{A}_2$ captures the u_1^2 term which did not appear until A_3 with the previous polynomials. (The $\hat{A}_n$ series converges faster.)

It is interesting to consider this further. Define $c_m^n=(u_m^n/n!)d^n/du_0^n$. First we can think of $c_m = u_m d/du_0$. Then $(c_m)(c_m) = (u_m d/du_0)(u_m d/du_0) = u_m^2 d^2/du_0^2$ (and if we then divide by the factorial of the number of repetitions, we have $c_m^2 = (u_m^2/2!)(d^2/du_0^2)$. Generalizing $c_m^n = (u_m^n/n!)d^n/du_0^n$. With this c_m^n consider the following rules:

$$A_0 = 1$$

$$A_1 = c_1$$

$$A_2 = c_2 + c_1^2$$

$$A_3 = c_3 + c_2c_1 + c_1^3$$

$$A_4 = c_4 + c_3c_1 + c_2(c_2 + c_1^2) + c_1^4$$

$$A_5 = c_5 + c_4c_1 + c_3(c_2 + c_1^2) + c_2(c_2c_1 + c_1^3) + c_1^5$$

$$A_6 = c_6 + c_5c_1 + c_4(c_2 + c_1^2) + c_3(c_3 + c_2c_1 + c_1^3)$$
$$+ c_2(c_2^2 + c_2c_1^2 + c_1^4) + c_1^6$$

$$A_7 = c_7 + c_6c_1 + c_5(c_2 + c_1^2) + c_4(c_3 + c_2c_1 + c_1^3)$$
$$+ c_3(c_3c_1^2 + c_2 + c_2c_1^2 + c_1^4)$$
$$+ c_2(c_2c_1^2 + c_2c_1^3 + c_1^5)\, c_1^7$$

$$A_8 = c_8 + c_7c_1 + c_6(c_2 + c_1^2) + c_5(c_3 + c_2c_1 + c_1^3)$$

$$+ c_4(c_4 + c_3c_1 + c_2^2 + c_2c_1^2 + c_1^4)$$
$$+ c_3(c_3c_2 + c_3c_1^2 + c_2^2c_1 + c_2c_1^3 + c_1^5)$$
$$+ c_2(c_2^3 + c_2^2c_1^2 + c_2c_1^4 + c_1^6) + c_1^8$$

$$A_9 = c_9 + c_8c_1 + c_7(c_2 + c_1^2) + c_6(c_3 + c_2c_1 + c_1^3)$$
$$+ c_5(c_4 + c_3c_1 + c_2^2 + c_2c_1^2 + c_1^4)$$
$$+ c_4(c_4c_1 + c_3c_2 + c_3c_1^2 + c_2^2c_1 + c_2c_1^3 + c_1^5)$$
$$+ c_3(c_3^2 + c_3c_2c_1 + c_3c_1^3 + c_2^3 + c_2^2c_1^2 + c_2c_1^4 + c_1^6)$$
$$+ c_2(c_2^3c_1 + c_2^2c_1^3 + c_2c_1^5 + c_1^7) + c_1^9$$

$$A_{10} = c_{10} + c_9c_1 + c_8(c_2 + c_1^2) + c_7(c_3 + c_2c_1 + c_1^3)$$
$$+ c_6(c_4 + c_3c_1 + c_2^2 + c_2c_1^2 + c_1^4)$$
$$+ c_5(c_5 + c_4c_1 + c_3c_2 + c_3c_1^2 + c_2^2c_1 + c_2c_1^3 + c_1^5)$$
$$+ c_4(c_4c_2 + c_4c_1^2 + c_3^2 + c_3c_2c_1$$
$$+ c_3c_1^3 + c_2^3 + c_2^2c_1^2 + c_2c_1^4 + c_1^6)$$
$$+ c_3(c_3^2c_1 + c_3c_2^2 + c_3c_2c_1^2 + c_3c_1^4$$
$$+ c_2^3c_1 + c_2^2c_1^3 + c_2c_1^5 + c_1^7)$$
$$+ c_2(c_2^4 + c_2^3c_1^2 + c_2^2c_1^4 + c_2c_1^6 + c_1^8) + c_1^{10}$$

Each A_m starts with c_m and ends with c_1^m. Successive terms after the first have c_{m-1}, c_{m-2}, ..., down to c_1. The second term is $c_{m-1}\, c_1$. Thus in the A_2 term since $m - 1 = 1$, the $c_{m-1}\, c_1$ is c^2. Clearly the sum of the subscripts is always m. In A_4 when we get to the c_2 term, either a c_2 or c_1^2 will preserve the sum m so we write $c_2(c_2 + c_1)$. In A_5 the next to the last term is $c_2(c_2c_1 + c_1^3)$ where we could conceivably have considered a c_3; however, we will use only subscripts less than or equal to the descending subscript (in the case c_2 or c_{m-3}). We can also indicate the procedure by writing

$$A_0 = 1$$

$$A_1 = c_1$$

$$A_2 = c_2 + c_1(c_1)$$

$$A_3 = c_3 + c_2(c_1) + c_1(c_1(c_1))$$

$$A_4 = c_4 + c_3(c_1) + c_2(c_2 + c_1(c_1)) + c_1(c_1(c_1(c_1)))$$

.
.
.

The result for this generating rule is exactly the same as for our previously given A_n. The only advantage, but an important one, is that no parameter λ is involved and only u_0 is needed which, of course, is physical, i.e., it contains the system input function as well as the auxiliary conditions. Now

$A_0 = f(u_0) = h(u_0)$ in our previous notation

$$A_1 = c_1 f(u_0) = u_1(d/du_0)f(u_0)$$

$$= u_1\ df(u_0)/du_0 = u_1 f^{(1)}(u_0) = u_1 h_1(u_0)$$

$$A_2 = (c_2 + c_1^2)f(u_0) = u_2 f^{(1)}(u_0) + (1/2!)\ u_1^2 f^{(2)}(u_0)$$

$$= u_2 h_1(u_0) + (1/2!)\ u_1^2 h_2(u_0)$$

$$A_3 = (c_3 + c_2 c_1 + c_1^3)f(u_0)$$

$$= u_3 f^{(1)}(u_0) + u_1 u_2\ f^{(2)}(u_0) + (1/3!)\ u_1^3\ f^{(3)}(u_0)$$

$$= u_3 h_1(u_0) + u_1 u_2 h_2(u_0) + (1/3!)\ u_1^3 h_3(u_0)$$

$$A_4 = u_4\ f^{(1)}(u_0) + [u_1 u_3 + (1/2!)\ u_2^2]\ f^{(2)}(u_0)$$

$$+\ u_2\ (1/2!)\ u_1^2\ f^{(3)}(u_0) + (1/4!)\ u_1^4\ f^{(4)}(u_0)$$

$$= u_4 h_1(u_0) + [u_1 u_3 + (1/2!)\, u_2^2] h_2(u_0) + (1/2!)\, u_1^2 u_2 h_3(u_0)$$

$$+ (1/4!)\, u_1^4 h_4(u_0)$$

This is more satisfactory since only the physical quantity $u_0(x)$ is involved. It is not, of course, a derivation. It is a generating rule for the A_n which allows decomposition to be successful.

Some final comparisons are worth mentioning. The $\hat{A}_n$ are easier to generate, do not involve the decomposition parameter λ, and lead to a more elegant derivation and a simpler proof of convergence, as well as much more rapid convergence. However, using the $\hat{A}_n$ to determine the solution, an algorithm for summation of the decomposition series for the solution is more difficult and integrability may become a problem in solving differential equations (since L^{-1} acts on $\Sigma\, \hat{A}_n$). The latter two points favor the A_n; the convergence is slower but still sufficiently rapid with the A_n, and a variety of generation formulas are possible. It seems clear that further generating formulas remain to be discovered which we leave to the reader. The price for the acceleration of convergence in possibly decreased integrability is analogous to the use of the entire linear term instead of only the highest ordered derivative for the L operator. Obviously the use of the former would accelerate convergence, but integration would be more difficult.

1.9 Euler's Transformation

Euler's transformation is sometimes a useful corollary to the decomposition method. We will use it to convert series solutions into more rapidly convergent, or, possibly finite series with an expanded radius of convergence, and, in some cases, an analytic solution. Consider the series

$$f(z) = \sum_{n=0}^{\infty} (-1)^n a_n z^n$$

If we multiply and divide by $1 + z$, we have

$$f(z) = (1/1+z) \left[\sum_{n=0}^{\infty} (-1)^n a_n z^n + z \sum_{n=0}^{\infty} (-1)^n a_n z^n \right]$$

$$= (1/1+z)\,[a_0 + \sum_{n=0}^{\infty} (-1)^{n+1} a_{n+1} z^{n+1}$$

$$+ \sum_{n=0}^{\infty} (-1)^n a_n z^{n+1}]$$

$$= (1/1+z)\,[a_0 + \sum_{n=0}^{\infty} (-1)^n (-a_{n+1})z^{n+1}$$

$$+ \sum_{n=0}^{\infty} (-1)^n a_n z^{n+1}]$$

$$= (1/1+z)\,[a_0 + \sum_{n=0}^{\infty} (-1)^n (a_n - a_{n+1})z^{n+1}]$$

$$= (a_0/1+z) + (z/1+z) \sum_{n=0}^{\infty} (-1)^n (\delta a_n)z^n$$

where $\delta a_n = a_n - a_{n+1}$. Write $f^{(1)}(z) = \sum_{n=0}^{\infty} (-1)^n (\delta a_n)z^n$. Thus $f(z) = (a_0/1+z) + (z/1+z)\, f^{(1)}(z)$. Suppose we now repeat the process on the series $f^{(1)}(z)$. Thus

$$f^{(1)}(z) = \sum_{n=0}^{\infty} (-1)^n (\delta a_n)z^n$$

$$= (1/1+z)[\sum_{n=0}^{\infty} (-1)^n (\delta a_n)z^n + z \sum_{n=0}^{\infty} (-1)^n (\delta a_n)z^n]$$

$$= (1/1+z)\,[\delta a_0 + \sum_{n+1=1}^{\infty} (-1)^{n+1}(\delta a_{n+1})z^{n+1}$$

$$+ \sum_{n=0}^{\infty} (-1)^n (\delta a_n)z^{n+1}]$$

$$= (1/1+z)\,[\delta a_0 + \sum_{n=0}^{\infty} (-1)^n (-\delta a_{n+1})z^{n+1}$$

$$+ \sum_{n=0}^{\infty} (-1)^n (\delta a_n)z^{n+1}]$$

$$= (1/1+z)\,[\delta a_0 + \sum_{n=0}^{\infty} (-1)^n (\delta a_n - \delta a_{n+1})z^{n+1}]$$

$$= (\delta a_0/1+z) + (z/1+z) \sum_{n=0}^{\infty} (-1)^n (\delta^{(2)} a_n)z^n$$

where $\delta^2 a_n = \delta a_n - \delta\, a_{n+1}$. Thus letting $f^{(2)}(z) = \sum_{n=0}^{\infty} (-1)^n(\delta^{(2)} a_n)z_n$ we have

$$f^{(1)}z = \delta a_0/1+z + (z/1+z)f^{(2)}(z)$$

Repeating the procedure, we get successively

$$f^{(2)}(z) = \delta^2 a_0/1+z + (z/1+z)f^{(3)}(z)$$

where $\delta^{(3)}a_n = \delta^{(2)}a_n - \delta^{(2)}a_{n+1}$ and $f^{(3)}(z) = \sum_{n=0}^{\infty} (-1)^n(\delta^{(3)} a_n)z^n$ and

$$f^{(3)}(z) = (\delta^3 a_0/1+z) + (z/1+z)f^{(4)}(z)$$

where $\delta^4 a_n = \delta^3 a_n - \delta^3 a_{n+1}$ and $f^{(4)}(z) = \sum_{n=0}^{\infty} (-1)^n(\delta^4 a_n)z_n$ etc. Consequently we write

$$f^{(m)}(z) \sum_{n=0}^{\infty} (-1)^n(\delta^m a_n)z^n = \delta^m a_0/1+z + (z+1/z)f^{(m+1)}(z)$$

Substituting $f^{(2)}(z)$, $f^{(3)}(z)$, ... into the expression for $f(z)$, e.g.

$$f(z) = a_0/1+z + (z/1+z)f^{(1)}(z) = a_0/1+z$$

$$+ (z/1+z)\{(\delta a_0/1+z) + (z/1+z)f^{(2)}(z)\} = \cdots$$

we have

$$f(z) = \sum_{n=0}^{m} (\delta^n a_0/1+z)(z/1+z)^n + (z/1+z)^{m+1} f^{(m+1)}(z)$$

where

$$f^{(m+1)}(z) = \sum_{n=0}^{\infty} (-1)^n (\delta^{m+1} a_n)z^n$$

Now consider the limit for high-order index m:

$$\lim_{m\to\infty} f(z) = \lim_{m\to\infty} \{\sum_{n=0}^{\infty} (\delta^n a_0/1+z)(z/1+z)^n$$

$$+ (z/z+1)^{m+1} \sum_{n=0}^{\infty} (-1)^n (\delta^{m+1} a_n)z^n \}$$

$$= \sum_{n=0}^{\infty} (\delta^n a_0/1+z)(z/1+z)^n$$

since for $|z| < \infty$, $|z/1+z| < 1$ and $\lim_{m\to\infty} (z/1+z)^{m+1} = 0$

The original series $f(z)$ diverges at $z = -1$, i.e., it has a radius of convergence $|z| < 1$ and in the limit $m\to\infty$,

$$f(z) = \sum_{n=0}^{\infty} (\delta^n a_0/1+z) + (z/1+z)^n$$

converges for finite z.

1.9.1 Solution of a Differential Equation by Decomposition

Let us consider the example: $dy/dx + e^y = 0$ with $y(0) = 1$ which is in our standard form (1983; 1986) $Ly + Ny = g(x)$ where $L = d/dx$, $Ny = e^y$, $g(x) = 0$ with the inverse L^{-1} defined as the definite integral from 0 to x.

$$L^{-1} Ly = - L^{-1} Ny = y - y(0)$$

$$y = y(0) - L^{-1} Ny$$

Letting $y = \sum_{n=0}^{\infty} y_n$ and $Ny = \sum_{n=0}^{\infty} A_n(y_0, \ldots, y_n)$

$$y_0 = y(0) = 1$$

$$y_{m\geq 1} = - L^{-1} A_{m-1}$$

The A_n are given by:

$$A_0 = e^{y_0}$$

$$A_1 = e^{y_0} (y_1)$$

$$A_2 = e^{y_0} (y_2 + y_1^2/2)$$

$$A_3 = e^{y_0} \{y_3 + y_1 y_2 + y_1^3/6\}$$

$$A_4 = e^{y_0} \{y_4 + y_2^2/2 + y_1 y_3 + y_1^2 y_2/2 + y_1^4/24\}$$

.
.
.

from which

$$y_0 = 1$$

$$y_1 = -\int_0^x e y_0 ds = -ex$$

$$y_2 = -\int_0^x e y_0 y_1 \, ds = (ex)^2/2$$

$$y_3 = -\int_0^x e y_0 \{y_2 + y_1^2/2\} ds = -(ex)^3/3$$

$$\vdots$$

$$y_m = -\int_0^x A_{m-1} ds = -(ex)^m/m$$

or

$$y = \sum_{n=0}^{\infty} y_n = 1 + \sum_{n=1}^{\infty} (-ex)^n/n$$

for which the radius of convergence is $1/e$.

1.9.2 Application of Euler Transform to Decomposition Solution

Let $ex = z$ and put the above series in the form $f(z)$. We have

$$y = 1 - \{z \sum_{m=1}^{\infty} (-1)^{m-1} z^{m-1}/m\}$$

$$= 1 - z \sum_{m+1=1}^{\infty} (-1)^{(m+1)-1} z^{(m+1)-1}/(m+1)$$

$$= 1 - z \{\sum_{n=0}^{\infty} (-1)^n (1/n+1) z^n\}$$

$$y = 1 - zh(z)$$

where

$$h(z) = \sum_{n=0}^{\infty} (-1)^n a_n z^n$$

with

$$a_n = 1/(n+1)$$

Then

$$h(z) = \sum_{n=0}^{\infty} (\delta^n a_0/1+z)\,(z/1+z)^n$$

Computing the coefficients $\delta^n a_0$ from $a_n = 1/(n+1)$,

n	a_n	δa_n	$\delta^2 a_n$	$\delta^3 a_n$	...
0	1	1/2	1/3	1/4	
1	1/2	1/6	1/12	1/20	
2	1/3	1/12	1/30		
3	1/4	1/20			
n	1/(n+1)				

Thus $\delta^n a_0 = 1/(n+1)$. Now

$$h(z) = (1/1+z) \sum_{n=0}^{\infty} (1/n+1)(z/1+z)^n$$

$$y = 1 - zf(z)$$

$$y = 1 - (z/1+z) \sum_{n=0}^{\infty} (1/n+1)(z/1+z)^n$$

$$y = 1 - \sum_{n=0}^{\infty} (1/n+1)(z/1+z)^{n+1}$$

Substituting the value for z we now have the transformed series

$$y = 1 - \sum_{n=0}^{\infty} (ex/1+ex)^{n+1}/(n+1) \qquad x < \infty$$

in comparison with the original series:

$$y = 1 + \sum_{n=1}^{\infty} (-ex)^n/n$$

$$= 1 + \sum_{n=0}^{\infty} (-1)^{n+1} (ex)^{n+1}/(n+1) \qquad |x| < 1/e$$

where we have used the subscripts to distinguish the two series for y. Taking the limit as $n \to \infty$ of $|y_{n+1}/y_n| < 1$ we verify the regions of convergence as stated above. Not only does the transformed decomposition series converge faster, but the circle of convergence is enlarged from radius $1/e$ to infinity.

Since the analytic sum is $y = 1 - ln(1 + ex)$ either

$$-\Sigma(-1)^{n+1}(1/1+n)(ex)^{n+1}$$

and

$$\Sigma(1/1+n)(ex/1+ex)^{n+1}$$

are expressions for $ln(1 + ex)$.

1.9.3 Numerical Comparison

We have

$$ln(1 + z) = z(1 - z/2 + z^2/3 - z^3/4 + \cdots)$$

with a radius of convergence $\rho = 1$. This series converges very slowly as $z \to 1$. The transformed series is

$$ln(1 + z) = (z/1+z)(1 + (1/2)(z/1+z) + (1/3)(z/1+z)^2 + \cdots$$

or

$$ln(1 + z) = -\sum_{n=1}^{\infty} (-1)^n z^n/n$$

$$ln(1 + z) = \sum_{n=1}^{\infty} (1/n)(z/1+z)^n$$

Let us compare by computing $ln\ 2$ using both series. We will use six terms, i.e., $\phi_6 = \sum_{n=0}^{5} y_n$ thus, letting $z = 1$ in the original series

$$\ln 2 = 1 - 1/2 + 1/3 - 1/4 + 1/5 - 1/6 + \cdots$$

$$= .616666 \quad \text{for six terms}$$

(an 11% error - which we expected since we took $z = 1$).

Now using the transformed decomposition series

$$\ln 2 = 1/2 + (1/2)(1/2)^2 + (1/3)(1/2)^3$$

$$+ (1/4)(1/2)^4 + (1/5)(1/2)^5 + (1/6)(1/2)^6 + \cdots$$

$$= .6911458 \quad \text{for six terms}$$

which is only an 0.29% error - less than 3/10 of a percent with six terms.

We see that the decomposition method is effective for solution, and we see the transformed decomposition series offers us a practical technique for a rapidly convergent analytic continuation. Let us compare with some common methods of solution for the same equation:

$$dy/dx + e^y = 0 \qquad y(0) = 1$$

Using separation of variables

$$dy/dx = -e^y$$

$$-e^{-y}\,dy = dx$$

$$\int_0^y e^{-y}(-dy) = \int_0^x dx$$

$$e^{-y} - e^{-y(0)} = x$$

$$e^{-y} - e^{-1} = x$$

$$e^{-y} = x + e^{-1}$$

Now multiplying by e

$$ee^{-y} = 1 + ex$$

$$ln\ e^{1-y} = ln\,(1 + ex)$$

$$1 - y = ln\,(1 + ex)$$

$$y = 1 - ln\,(1 + ex)$$

which verifies the previous solution. Remember that the previous method is much more widely applicable, but we have used an equation which could also be solved by usual methods for the sake of comparison.

1.9.4 Solution of Linearized Equation

Write $e^y = 1 + y$, i.e., we assume small y . Now the differential equation becomes

$$dy/dx + y + 1 = 0 \qquad\qquad y(0) = 1$$

$$Ly = -1 - y$$

$$L^{-1}Ly = y - y(0) = y - 1$$

$$y = 1 - L^{-1}(1) - L^{-1}y$$

$$y = 1 - x - L^{-1}\sum_{n=0}^{\infty} y_n$$

We have, therefore,

$$y_0 = 1 - x$$

$$y_1 = -L^{-1}y_0 = -x + x^2/2!$$

$$y_2 = -L^{-1}\ y_1 = x^2/2! - x^3/3!$$

$$y_3 = -L^{-1}y_2 = -x^3/3! + x^4/4!$$

.
.
.

$$y_n = (-1)^n x^n/n! + (-1)^{n+1} x^{n+1}/(n+1)!$$

Hence

$$y = \sum_{n=0}^{\infty} \{(-1)^n x^n/n! + (-1)^{n+1} x^{n+1}/(n+1)!\}$$

$$= (e^{-x}) + (e^{-x} - 1) = 2e^{-x} - 1$$

for $|x| < \infty$ but the problem has changed! Thus transformations such as the Euler appear more valuable now that the decomposition is solving the *nonlinear* equations in series form.

Note that in the Picard form, we have

$$dy/dx = -e^y \qquad y(0) = 1$$

and the Picard iterates are

$$\phi_0 = 1$$

$$\phi_1 = 1 + \int_0^y (-e^{\phi_0})dx = 1 - ex$$

$$\phi_2 = 1 + \int_0^x (-e^{1-ex})dx$$

$$= 1 - ex + e^{-1}\{e^{-ex} - 1\}$$

$$\phi_3 = 1 + \int_0^x (-e^{\phi_2(x)})dx$$

which is clearly not a way to proceed, again emphasizing that there is no connection between Picard schemes and decomposition solutions. Let us consider another example:

Example:

$$dy/dx + x^{m-1}\, y^2 = 0 \qquad\qquad y(0) = m$$

We know the solution in advance since differentiating $y = m(1 + x^m)^{-1}$ yields $dy/dx = -\ x^{m-1}\, m^2(1 + x^m)^{-2} = x^{m-1}y^2$. Thus the differential equation is satisfied. Our objective is to show that for a class of nonlinear differential equations (which we leave to future dissertations to define), Euler transformation of the decomposition solution by our methods can lead to a finite series or even the sum function, i.e., an exact solution instead of an "approximation". Write $L = d/dx$, $L^{-1} = \int_0^x [\cdot]\, dx$, $Ny = y^2 = \sum_{n=0}^{\infty} A_n$ and $y = \sum_{n=0}^{\infty} y_n$. Now we have

$$Ly = -\ x^{m-1}\, Ny$$

$$L^{-1}\, Ly = -\ L^{-1}\, x^{m-1} Ny$$

$$y = y(0) - L^{-1}\, x^{m-1} Ny$$

$$y = y_0 - L^{-1}\, x^{m-1} \sum_{n=0}^{\infty} A_n$$

Given now that

$$A_0 = y_0^2$$

$$A_1 = 2y_0y_1$$

$$A_2 = y_1^2 + 2y_0y_2$$

$$A_3 = 2y_1y_2 + 2y_0y_3$$

etc. for A_3, A_4, ... see [1983; 1986], we have the components of y

$$y_0 = m$$

$$y_1 = -\ L^{-1}\, x^{m-1}\, A_0 = -\ L^{-1}\, x^{m-1}\, m^2 = -mx^m$$

$$y_2 = -\ L^{-1}\, x^{m-1}\, A_1 = -\ L^{-1}\, x^{m-1}\, 2m(-mx^m) = mx^{2m}$$

$$y_3 = -L^{-1} x^{m-1} \{m^2x^{2m} + 2m(mx^m)\} = -mx^{3m}$$

.

.

.

$$y_n = (-1)^n mx^{nm}$$

We now have the solution of the differential equation as

$$y = m \sum_{n=0}^{\infty} (-1)^n (x^m)^n$$

whose analytic sum, i.e., the exact or closed form is

$$y = m/(1 + x^m) \qquad (x^m \neq -1)$$

Since we have the form $f(z) = \sum_{n=0}^{\infty} (-1)^n a_n z^n$ and can Euler-transform[3] into the form $f(z) = (1/1+z) \sum_{n=0} (\delta^n a_0)(z/1 + z)^n$ where δ is the forward difference operator such that

$$\delta a_n = a_n - a_{n+1},$$

$$\delta^2 a_n = a_n - 2a_{n+1} + a_{n+2},$$

$$\delta^3 a_n = a_n - 3a_{n+1} + 3a_{n+2} - a_{n+3}, \ldots,$$

or

$$\delta^k a_n \sum_{p=0}^{k} (-1)^p \binom{k}{p} a_{n+p},$$

we make the identification of our solution with the Euler form by $a_n = m$ and $z = x^m$ obtaining

n	a_n	δa_n	$\delta^2 a_n$	$\delta^3 a_n$	...
0	m	0	0	0	
1	m	0	0	0	
2	m	0	0	0	
3	m	0	0	0	
4	m	0	0	0	
.	.				
.	.				
	.				

[3]There exist more general formulas, but the given form is convenient here.

Thus $\delta^0 a_0$ and $\delta^{n\geq 1} a_0 \equiv 0$. The Euler-transformed series is therefore given by

$$y = 1/(1+z) \sum_{n=0}^{\infty} (\delta^n a_0)(z/1+z)^n = a_0/(1+z) = m/(1+x^m)$$

This particular solution was obvious, of course, but it illustrates the method again. The convergence region is not increased here because we have an intrinsic singularity at -1.

Thus the decomposition series solution becomes the exact solution. (It diverges only at $x = -1$) if m is odd and at $\pm i$ for m even.) Clearly then, the Euler transformation can be a useful tool for a class of nonlinear differential equations not only to obtain more rapid acceleration but also to obtain a finite series or even an exact solution from the decomposition series.

What about other acceleration schemes? A useful scheme was discussed in (1983). There, we could compute an n-term approximation and calculate what we called a "correction" term which allowed us to see what the effect of more terms would be, to see if it was worth going further with the computation. It seems obvious that many other methods can be developed using complex analysis. Or, we can find the effect of the above-mentioned "correction" on an Euler-transformed series. One final method[4] we will discuss here is the following. Consider

$$S(x) = \sum_{n=1}^{\infty} (-1)^n (n \cos nx)/(n^3 + 7) \qquad 0 < x < \pi$$

The absolute value of the general term is less than $1/n^2$ which converges, hence the series converges. Call the sum $S(x)$. For large n, $(-1)^n/n^2$ is a good approximation for the coefficients so that we can write

$$\sum_{n=1}^{\infty} (-1)^n (\cos nx)/n^2 \qquad -\pi \leq x \leq \pi$$

But this is the series we can get by writing the Fourier series for x^2

$$x^2 = (\pi^2/3) + 4 \sum_{n=1}^{\infty} (-1)^n (\cos nx)/n^2$$

---

[4]From a course by Professor Earl D. Rainville at the University of Michigan a long time ago...

so that

$$\sum_{n=1}^{\infty} (-1)^n (\cos nx)/n^2 = (1/4)(x^2 - \pi^2/3)$$

Since the coefficients in the series whose sum is known and the series $S(x)$ are nearly equal as n gets large, the differences should be small again. Thus

$$S(x) - (1/4)(x^2 - \pi^2/3) = \sum_{n=1}^{\infty} (-1)^n [(n/n^3+7) - 1/n^2)] \cos nx$$

$$= \sum_{n=1}^{\infty} (7(-1)^{n+1} \cos nx)/(n^2(n^3+7))$$

or

$$S(x) = (1/4)(x^2 - \pi^2/3) + 7 \sum_{n=1}^{\infty} ((-1)^{n+1} \cos nx)/(n^2(n^3+7))$$

for $0 < x < \pi$. This series converges much faster since coefficients of $\cos nx$ get smaller faster. The key here was knowing the sum of the comparison series and using the differencing idea again as in the Euler transformation.

A question still open is the consideration of similar acceleration schemes for decomposition solutions of nonlinear partial differential equations. We have seen here that accelerated convergence and analytic continuation into a widened convergence region is achieved by Euler-transforming the series solution. The global applicability and utility of the decomposition method for nonlinear equations gives a new importance to the Euler transformation and other acceleration schemes.

1.10 On the Validity of the Decomposition Solution

Suppose we consider a linear equation $Fu = Lu + Ru = \xi$ where $\xi = \xi(t)$ for $t \geq 0$, L is a simple differential operator, and R is the remainder operator $F - L$. Write $Lu = \xi - Ru$ and operate on both sides with L^{-1} to write

$$u = u_0 - Ku \qquad (1.10.1)$$

where $K = L^{-1}R$ and $u_0 = \phi + L^{-1}\xi$ where $L\phi = 0$. We can write now

$$(I + K)u = u_0 \tag{1.10.2}$$

The decomposition solution is obtained immediately in our usual way as

$$u = \sum_{n=0}^{\infty} (-1)^n K^n u_0 \tag{1.10.3}$$

but we observe immediately that substitution of (1.10.3) into (1.10.2) is an identity. Or, equivalently, from (1.10.1), writing $u = \sum_{n=0} u_n$ and identifying $u_{n+1} = -Ku_n$ for $n \geq 0$, we again have (1.10.3).

To generalize to the equation $(L_x + L_t)u + Ru = \xi$ where L_x and L_t are differentiations with respect to x and t respectively, we operate on both sides with $(L_x^{-1} + L_t^{-1})$. Then $L_t^{-1} L_t u = u - \phi_t$ and $L_x^{-1} L_x u = u - \phi_x$, and we obtain immediately

$$u = u_0 - Ku \tag{1.10.4}$$

where

$$u_0 = (1/2)(\phi_t + \phi_x) + (1/2)(L_t^{-1} + L_x^{-1})\xi$$

$$K = (1/2)(L_x^{-1} L_t + L_t^{-1} L_x) - (1/2)(L_t^{-1} + L_x^{-1})R$$

and we see (1.10.4) is also the decomposition solution if we write $u = \sum_{n=0}^{\infty} u_n$ and identify $u_{n+1} = -Ku_n$ for $n \geq 0$. When $R = 0$, $(L_t + L_x)u = \xi$ is solved by $\sum_{n=0}^{\infty} (-1)^n K^n u_0$.

If we consider the nonlinear case $Lu + Ru + Nu = \xi$, we write $Lu = \xi - Ru - Nu$, operate with L^{-1} to get $u = u_0 - Ku - L^{-1} Nu$ where $u_0 = \phi + L^{-1}\xi$ and $K = L^{-1}R$. Decomposition of u into $\sum_{n=0}^{\infty} u_n$ yields

$$u_{n+1} = Ku_n - L^{-1} A_n \qquad n \geq 0 \tag{1.10.5}$$

The two-dimensional case $(L_t + L_x)u + Ru + Nu = \xi$ yields $u_0 = (1/2)(\phi_t + \phi_x) + (1/2)(L_t^{-1} + L_x^{-1})\xi$ and $K = (1/2)(L_t^{-1} L_x + L_x^{-1} L_t)$ and the operator on A_n becomes $(1/2)(L_t^{-1} + L_x^{-1})$. For a linear operator $L_t + L_x + L_y + L_z$ (such as $(\partial/\partial t + \nabla^2)$ for example), the operator on A_n becomes $(1/4)(L_t^{-1} + L_x^{-1} + L_y^{-1} + L_z^{-1})$ and K becomes

$$(1/4)\{L_t^{-1}(L_x + L_y + L_z) + L_x^{-1}(L_y + L_z + L_t) + L_y^{-1}(L_z + L_t + L_x) + L_z^{-1}(L_t + L_x + L_y)\}.$$

Thus, the method is quite clear and demonstrably valid. We have simply decomposed u into components $\sum_{n=0}^{\infty} u_n$ and $Nu = \sum_{n=0}^{\infty} A_n$ and determined the components such that $u = \sum_{n=0}^{\infty} u_n$. The approximant is $\sum_{i=0}^{n-1} u_i$ whose limit is u. The $\sum_{n=0}^{\infty} A_n$ with A_n dependent only on u_0 through u_n makes the system calculable. We note that if the homogeneous solutions vanish in the linear equations, we can write $Fu = \xi$ as $u = F^{-1}\xi$ and that this is no longer only a formal solution but a useful one since F^{-1} is now determinable.

References

Adomian, G., *Nonlinear Stochastic Operator Equations* . Academic Press, 1986.

Adomian, G. "The Solution of General Linear and Nonlinear Sytems," *Modern Trends in Cybernetics and Systems* , ed. J. Rose, Editura Technica, Springer-Verlag, 1976.

Adomian, G., *Stochastic Systems* . Academic Press, New York, 1983.

Adomian, G., L. H. Sibul, and R. Rach, "Coupled Nonlinear Stochastic Differential.Equations," *J. Math. Anal. and Appl.* , 92, 2, 1983, 427-434.

Bellman, R. E. and G. Adomian. *Partial Differential Equations - New Methods for Their Treatment and Application.* Reidel, 1985.

Bellomo, N. and R. Monaco, "A Comparison between Adomian's Decomposition Method and Perturbation Techniques for Nonlinear Random Differential Equations," *J. Math. Anal. and Appl.* , 110, 1985, 495-502.

Bellomo, N. and R. Riganti, *Nonlinear Stochastic Systems in Physics and Mechanics* , World Scientific Publishers, 1987.

Rach, R., "A Convenient Computational Form for the Adomian Polynomials," *J. Math. Anal. and Appl.* 102, 2, 1984, 415-419.

Riganti, R., "Transient Behavior of Semilinear Stochastic Systems with Random Parameters," *J. Math. Anal. and Appl.* , 98, 1984, 314-327.

CHAPTER 2

Effects of Nonlinearity and Linearization

2.1 Introduction

Nonlinear equations arise in every area of application, and the correct solution of dynamical systems modeled by nonlinear ordinary differential equations, systems of differential equations, partial differential equations, and systems of partial differential equations is vital to progress in many fields. In order to make these equations tractable, it is quite common to linearize equations or assume "weak" nonlinearity, etc., because adequate methods simply have not been otherwise available. It is known, of course, that the linearized solution can deviate considerably from the actual solution of the nonlinear problem and that linearization procedures require proof that the solution is valid. For example, writing $\ddot{x} = a \sin x$ in the form $\ddot{x} = ax$ requires a priori proof that x is sufficiently small. The decomposition method (1983; 1986; 1985) has substantially improved our ability to solve a wide class of nonlinear and/or stochastic equations. For example, it is now possible to obtain very accurate and verifiable solutions of nonlinear, or even nonlinear stochastic, equations for all of the above types even if nonlinear, stochastic or coupled boundary conditions are involved (1986). Usage of linearization has become rather standardized; however, solution of the actual nonlinear form is clearly preferable to a linear approximation. Since the linearized problem is a different problem (as Casti has pointed out, 1985a), the usage of linearization requires justification that it is adequate in a particular problem to change it in this way. The practice of approximating a nonlinear function with a linearized version arose from the need to make equations tractable by simple analysis, since numerical solutions from computers have drawbacks and methods of analytical solution of nonlinear equations are generally inadequate. The belief that linearization and perturbation are essential procedures, the reluctance to give up the convenient analytical tool of superposition, and the faith that faster computers will solve everything are factors in preserving the status quo.

Exact linearization is possible for a nonlinear equation - as in the case of Burger's equation - so that a convenient check can in some cases be made of the decomposition solution. As an example, consider the nonlinear equation for $\phi(x,t)$

$$\phi_t + \phi_x + a\phi + \phi^2 = 0$$

with specified conditions. The transformation $\phi = 1/\psi(x,t)$ leads to the linear equation

$$\psi_t + \psi_x - \alpha\psi = 1$$

and conditions specified on ψ; however, the nonlinear equation can be solved directly as follows. Using decomposition (1983;1985;1986), we write

$$L_t\phi + L_x\phi + a\phi + \phi^2 = 0$$

Let $N\phi = \phi^2 = \sum_{n=0}^{\infty} A_n$ where

$$A_0 = \phi_0^2$$

$$A_1 = 2\phi_0\phi_1$$

$$A_2 = \phi_1^2 + 2\phi_0\phi_2$$

$$A_3 = 2\phi_1\phi_2 + 2\phi_0\phi_3$$

$$\vdots$$

Solving for $L_t\phi$ and for $L_x\phi$

$$L_t\phi = - L_x\phi - a\phi - \phi^2$$

$$L_x\phi = - L_t\phi - a\phi - \phi^2$$

Thus

$$\phi = \phi(x,0) - L_t^{-1} L_x\phi - a L_t^{-1} \phi - L_t^{-1} \phi^2$$

$$\phi = \phi(0,t) - L_x^{-1} L_t\phi - a L_x^{-1} \phi - L_x^{-1} \phi^2$$

Adding

$$\phi = (1/2)\{\phi(x,0) + \phi(0,t) - (L_t^{-1} L_x + L_x^{-1} L_t)\phi$$

$$- a(L_t^{-1} + L_x^{-1})\phi - (L_t^{-1} + L_x^{-1})\phi^2\}$$

We define

$$\phi_0 = (1/2)\{\phi(x,0) + \phi(0,t)\}$$

substitute $\phi = \sum_{n=0}^{\infty} \phi_n$ and $\phi^2 = \sum_{n=0}^{\infty} A_n$ to obtain

$$\phi_{n+1} = -(1/2)\{L_t^{-1} L_x + L_x^{-1} L_t\}\phi_n$$

$$-(1/2)a\{L_t^{-1} + L_x^{-1}\}\phi_n - (1/2)\{L_t^{-1} + L_x^{-1}\}A_n$$

for $n \geq 0$ so that all components are determined.

2.2 Effects on Simple Systems

In the analysis of systems it is common to suppose that everything is linear or sufficiently close to linear so that linearized analyses will be adequate. Thus in a mechanical system assumed to be linear, displacements and accelerations are proportional to forces. If the system is electrical, currents are proportional to voltages, etc. Thus we write output $y(t)$ as proportional to input $x(t)$ or $y = kx$ where k is a constant independent of t or x. Now suppose that the system deviates only slightly from linearity by adding a small term which is nonlinear, e. g., $y = kx + \varepsilon x^2$. An example from Professor Richard Feynman's (1963-65) lectures considers the input: $x = \cos \omega t$ and consequently in the linear system $y = k \cos \omega t$. In the slightly nonlinear case, we have

$$y = k[\cos \omega t + \varepsilon \cos^2 \omega t]$$

$$= k[\cos \omega t + (\varepsilon/2)(1 + \cos 2\omega t)]$$

so that in addition to the $\cos \omega t$ term, we get a second harmonic term and a constant term $k\varepsilon/2$ corresponding to a shift of the average value. A nonlinear term such as x^3 or x^4 will produce still higher harmonics. If we begin with a general nonlinear term $f(x)$ and linearize it, we obviously lose such terms.

Now consider the input $x = A\cos\omega_1 t + B\cos\omega_2 t$ which yields not only the linear term

$$k[A\cos\omega_1 t + B\cos\omega_2 t]$$

but also

$$k\varepsilon[A^2\cos^2\omega_1 t + B^2\cos^2\omega_2 t + 2AB\cos\omega_1 t\cos\omega_2 t]$$

The first two of these produce constant terms and second harmonic terms as before. Also sum and difference frequencies arise from the cross product term

$$2AB\cos\omega_1 t\cos\omega_2 t = AB[\cos(\omega_1 + \omega_2)t + \cos(\omega_1 - \omega_2)t]$$

Thus, a nonlinear system produces new effects not present in linear systems; these effects are proportional to ε (and to products of amplitudes A^2, B^2, or AB). Clearly then, if ε is not small, such effects become important.

Let us consider a more complicated linear example - the well-known harmonic oscillator equation $\ddot{x} + (k/m)x = F/m$. Let $F = F_0\exp\{i\omega t\}$ and let $\omega_0^2 = k/m$. Since the solution will have the same frequency as the applied force, we can write $x = x_0\exp\{i\omega t\}$. Substituting in the differential equation, we get

$$-\omega^2 x_0 + \omega_0^2 x_0 = F_0/m$$

$$(\omega_0^2 - \omega^2)x_0 = F_0/m$$

$$x_0 = (F_0/m)/(\omega_0^2 - \omega^2)$$

If ω approaches ω_0, x_0 becomes very large and theoretically approaches infinity. Remember, however, that friction has been neglected. The frictional force is usually proportional to the velocity and can be written conveniently as $-\Upsilon m\dot{x}$ where Υ is a constant. Then we have

$$\ddot{x} + \Upsilon\dot{x} + \omega^2 x = F/m$$

and it is easily found that the amplitude of oscillation is now $x_0 = (F_0/m)(\omega_0^2 - \omega^2 + i\Upsilon\omega)$ and now the peak or resonance value of x_0 is finite. Since $(1/m)(\omega_0^2 - \omega^2 + i\Upsilon\phi)$ is complex, let it equal $\rho e^{i\phi}$. Then $x_0 = F_0\rho e^{i\phi}$ and the magnitude of the oscillation is seen to be $F_0\rho$, but the oscillation is no longer in phase with F_0 but shifted in phase by ϕ, and it is easily demonstrated that ϕ satisfies $\tan\phi = -\Upsilon\omega/(\omega_0^2 - \omega^2)$.

The assumption that the frictional force is simply proportional to the velocity $\dot{x}$ does not always hold; the constant Υ may be velocity-dependent so that we get a nonlinear term. We may have a small additional term proportional to $\dot{x}^2$, for example, and get

$$\ddot{x} + \Upsilon\dot{x} + \omega^2 x + \varepsilon\dot{x}^2 = F/m$$

which is nonlinear. The solution will then be generally carried out only under the assumption that ε is small so that perturbation theory will be applicable, i.e., when we consider a "slightly nonlinear" or a "weakly nonlinear" system.

Exercise:

Solve the weakly nonlinear case by perturbation theory and consider the effects of the small nonlinearity.

2.3 Effects on Solution for the General Case

Since, in general, solutions of nonlinear equations are made by linearizing the equations, it is natural to ask what the effect of linearization is on the actual solutions. Let us consider the equation

$$Ly + Ry + Ny = x(t)$$

where L is the invertible linear operator, R is the remaining linear operator, and Ny is the nonlinear term. We have

$$Ly = x - Ry - Ny$$

$$y = \Phi + L^{-1}x - L^{-1}Ry - L^{-1}Ny$$

where $L\Phi = 0$. We assume the solution decomposition $y = \sum_{n=0}^{\infty} y_n$

with $y_0 = \Phi + L^{-1} x$ and also the decomposition of the nonlinear term Ny into $\sum_{n=0}^{\infty} A_n$ where the A_n are generated for the specific Ny. Then the components after y_0 are determinable in terms of y_0 as

$$y_1 = - L^{-1} \; Ry_0 - L^{-1} \; A_0(y_0)$$

$$y_2 = - L^{-1} \; Ry_1 - L^{-1} \; A_1(y_0,y_1)$$

$$y_3 = - L^{-1} \; Ry_2 - L^{-1} \; A_2(y_0,y_1,y_2)$$

$$\vdots$$

$$y_n = - L^{-1} Ry_{n-1} - L^{-1} A_{n-1}(y_0,\ldots,y_{n-1})$$

or, equivalently,

$$y_1 = (- L^{-1}R)y_0 - L^{-1} A_0$$

$$y_2 = (- L^{-1}R)^2 y_0 - (- L^{-1}R) L^{-1} A_0 - L^{-1} A_1$$

$$y_3 = (- L^{-1}R)^3 y_0 - (- L^{-1}R)^2 L^{-1} A_0 - (- L^{-1}R)L^{-1} A_1 - L^{-1} A_2$$

$$\vdots$$

$$y_n = (- L^{-1}R)^n y_0 - \sum_{\upsilon=0}^{n-1} (- L^{-1}R)^{n-1-\upsilon} L^{-1} A_\upsilon$$

for $n \geq 1$. The solution is $y = \sum_{n=0}^{\infty} y_n$ or

$$\begin{aligned} y &= \sum_{n=0}^{\infty}(- L^{-1}R)^n y_0 \\ &= \sum_{n=1}^{\infty}\sum_{\upsilon=0}^{n-1}(- L^{-1}R)^{n-1-\upsilon} - L^{-1}A_\upsilon \end{aligned} \qquad (2.3.1.)$$

We leave it to the reader as a simple exercise to show that $A_n(y_0,y_1,\ldots,y_n)$ reduces to y_n if $Ny = f(y) = y$. Then the solution is

$$y = \sum_{n=0}^{\infty} (-L^{-1}R)^n y_0 - \sum_{n=1}^{\infty}\sum_{\upsilon=0}^{n-1}(-L^{-1}R)^{n-1-\upsilon} L^{-1} y_\upsilon \qquad (2.3.2)$$

i.e., the solution corresponds now to the equation $Ly + Ry + y = x$ which yields

$$y = y_0 - L^{-1}Ry - L^{-1} y$$

with

$$y_1 = -L^{-1} Ry_0 - L^{-1} y_0$$

$$y_2 = -L^{-1} Ry_1 - L^{-1} y_0$$

$$= (-L^{-1} R)^2 y_0 - (-L^{-1}R) L^{-1} y_0 - L^{-1} y_0$$

$$y_3 = -L^{-1} Ry_2 - L^{-1} y_1$$

$$= (-L^{-1} R)^3 y_0 - (-L^{-1}R)^2 L^{-1} y_0 - (-L^{-1} R) L^{-1} y_0 - L^{-1}y_1$$

$$\vdots$$

The result on the solution of replacing $Ny = f(y)$ by y is seen by plotting (2.3.1) and (2.3.2). Similarly replacing $f(y)$ by another perhaps more sophisticated linearization is seen by simply using (2.3.1) and calculating the A_n for the linearized function replacing $f(y)$. For a general linear stochastic system it has been shown[1] that the decomposition solution reduces to the results of perturbation theory in cases where perturbation theory is applicable; however, the solution is not restricted to "small" fluctuations as in the perturbation result. This is also true of nonlinear stochastic or nonlinear deterministic systems; the methods proposed involve no "small fluctuation" or "small nonlinearity" assumptions.

---

[1]Adomian, G., "Linear Random Operator Equations in Math. Phys., II," *J. Math. Phys.*, 12, 9, 1971, 1944-1948.

Example:

Exponential nonlinearity: Let us consider a simple nonlinear ordinary differential equation with an exponential nonlinearity

$$dy/dx + e^y = 0$$

$$y(0) = 1$$

In our usual standard form (1983) this is written

$$Ly + Ny = x$$

with $L = d/dx$ and $Ny = e^y$. We solve for Ly, i.e., $Ly = -Ny$ then write $L^{-1}Ly = -L^{-1}Ny$ with L^{-1} defined as the integration over x. Thus

$$y = y(0) - L^{-1}Ny$$

The nonlinear term $Ny = e^y$ is replaced by $\sum_{n=0}^{\infty} A_n$ where the A_n can be written as $A_n(e^y)$ to emphasize that they are generated for this specific function. Thus,

$$y = y(0) - L^{-1}\sum_{n=0}^{\infty} A_n(e^y)$$

Now the decomposition of the solution y into $\sum_{n=0}^{\infty} y_n$ leads to the term-by-term identification

$$y_0 = y(0) = 1$$

$$y_1 = -L^{-1}A_0$$

$$y_2 = -L^{-1}A_1$$

$$\vdots$$

$$y_{n+1} = -L^{-1}A_n$$

The $A_n(e^y)$ are given by (1983; 1986):

$$A_0 = e^{y_0}$$

$$A_1 = y_1 e^{y_0}$$

$$A_2 = (y_1^2/2 + y_2)e^{y_0}$$

$$A_3 = (y_1^3/6 + y_1 y_2 + y_3)e^{y_0}$$

.
.
.

Thus

$$y_0 = 1$$

$$y_1 = -ex$$

$$y_2 = e^2x^2/2$$

$$y_3 = -e^3x^3/3!$$

.
.
.

$$y = 1 + \sum_{n=1}^{\infty} (-1)^n e^n x^n/n!$$

which can also be written $y = 1 - \ln[1 + ex]$ for $x < 1$.

Suppose we replace e^y by $1 + y$, dropping all terms of the series for e^y except the constant and the linear term. The differential equation becomes

$$dy/dx = -(1 + y)$$

Then

$$y = y(0) - L^{-1}(1 + y)$$

$$= 1 - L^{-1}[1] - L^{-1}\sum_{n=0}^{\infty} y_n$$

so that

$$y_0 = 1 - x$$

$$y_1 = -L^{-1}(1 - x)$$

$$\vdots$$

$$y_n = -L^{-1} y_{n-1} \qquad (n \geq 1)$$

Since y is the sum of the components, we have

$$y = \sum_{n=0}^{\infty} \{(-1)^n x^n/n! + (-1)^{n+1} x^{n+1}/(n+1)!\}$$

$$y = e^{-x} + (e^{-x} - 1) = 2e^{-x} - 1$$

We will identify this linearized solution as y_ℓ and compare with the solution y of the nonlinear equation. The results are given in the table

x	y	y_ℓ
0	1.0000	1.0000
1	. 7595	.8097
.2	. 5658	.6375
3	.4036	.4816
.4	.2641	.3406
.5	.1417	.2131

Example:

Anharmonic oscillator: Consider now the anharmonic oscillator $d^2\theta/dt^2 + k^2 \sin\theta = 0$ for $\theta(0) = \Upsilon$ = constant and $\theta'(0) = 0$. Using the decomposition method, the solution is found to be

$$\theta(t) = \Upsilon - [(kt)^2/2!]\sin \Upsilon + [(kt)^4/4!]\sin \Upsilon \cos \Upsilon$$

$$- [(kt)^6/6!](\sin \Upsilon \cos^2 \Upsilon - 3 \sin^3 \Upsilon] + \ldots$$

which becomes

$$\theta(t) - \Upsilon[1 - (k^2t^2/2!) + (kt)^4/4! - \cdots]$$

in the linearized case, i.e., for "small amplitude" motion which offers interesting comparison when the smallness assumption is inappropriate.

Example:

Hyperbolic sine nonlinearity: Consider the equation

$$du/dt - k \sinh (u/\alpha)$$

where $u(0) = c > 0$ for $t > 0$.

If we assume that we can approximate $\sinh u/\alpha \cong u/\alpha$, we have $du/dt - ku/\alpha = 0$. Now solving by decomposition with $u = \sum_{n=0}^{\infty} u_n$, $L = d/dt$, and L^{-1} as the definite integral from 0 to t

$$Lu - ku/\alpha = 0$$

$$L^{-1}Lu = L^{-1}(k/\alpha)u$$

$$u = u(0) + L^{-1}(k/\alpha)u = \sum_{n=0}^{\infty} u_n$$

$$u_0 = u(0) = c$$

$$u_1 = (k/\alpha)ct = ckt/\alpha$$

$$u_2 = c(kt/\alpha)^2/2!$$

$$u_3 = c(kt/\alpha)^3/3!$$

$$\vdots$$

$$u_m = c(kt/\alpha)^m/m!$$

$$\vdots$$

i.e., $u = ce^{kt/\alpha}$. To solve the original equation with $\sinh u/\alpha$

$$u = \sum_{n=0}^{\infty} u_n = c + k L^{-1} \sum_{n=0}^{\infty} A_n$$

where the A_n are generated for $Nu = \sinh u/\alpha$. These are given by:

$$A_0 = \sinh(u_0/\alpha)$$

$$A_1 = (u_1/\alpha) \cosh(u_0/\alpha)$$

$$A_2 = (u_2/\alpha) \cosh(u_0/\alpha) + (1/2!)(u_1^2/\alpha^2) \sinh(u_0/\alpha)$$

$$A_3 = (u_3/\alpha) \cosh(u_0/\alpha) + (u_1/\alpha)(u_2/\alpha) \sinh(u_0/\alpha)$$
$$+ (1/3!)(u_1^3/\alpha^3) \cosh(u_0/\alpha)$$

$$A_4 = (u_4/\alpha) \cosh(u_0/\alpha) + [(1/2!)(u_2^2/\alpha^2)$$
$$+ (u_1/\alpha)(u_2/\alpha)]\sinh(u_0/\alpha)$$
$$+ (1/2!)(u_1^2/\alpha^2)(u_2/\alpha) \cosh(u_0/\alpha)$$
$$+ (1/4!)(u_1^4\alpha^4) \sinh(u_0/\alpha)$$

$$A_5 = (u_5/\alpha) \cosh(u_0/\alpha) + [(u_2/\alpha)(u_3/\alpha)$$
$$+ (u_1/\alpha)(u_4/\alpha)] \sinh(u_0/\alpha)$$
$$+ [(u_1/\alpha)(1/2!)(u_2^2/\alpha^2)$$
$$+ (1/2!)(u_1^2/\alpha^2)(u_3/\alpha)] \cosh(u_0/\alpha)$$

$$+ (1/3!)(u_1^3/\alpha^3)u_2/\alpha) \sinh(u_0/\alpha)$$
$$+ (1/5!)(u_1^5/\alpha^5) \cosh(u_0/\alpha)$$

$$A_6 = (u_6\alpha) \cosh(u_0/\alpha) + [(1/2!)(u_2^3/\alpha^2) + (u_2/\alpha)u_4/\alpha)$$
$$+ (u_1/\alpha)(u_5/\alpha)] \sinh(u_0/\alpha) +$$

$$[(1/3!)(u_2^3/\alpha^3) + (u_1/\alpha)(u_2/\alpha)(u_3/\alpha)$$

$$+ (1/2!)(u_1^2/\alpha^2)(u_4/\alpha)] \cosh(u_0/\alpha)$$

$$+ [(1/2!)(u_1^2/\alpha^2)(1/2!)(u_2^2/\alpha^2)$$

$$+ (1/3!)(u_1^3/\alpha^3)(u_3/\alpha)] \sinh(u_0/\alpha)$$

$$+ (1/4!)(u_1^4/\alpha^4)(u_2/\alpha) \cosh(u_0/\alpha)$$

$$+ (1/6!)(u_1^6/\alpha^6) \sinh(u_0/\alpha)$$

Now

$$u_0 = c$$

$$u_1 = kL^{-1} A_0 = kL^{-1}[\sinh u_0/\alpha] = kt \sinh c/\alpha$$

$$u_2 = kL^{-1} A_1 = kL^{-1}[(u_1/\alpha) \cosh(u_0/\alpha)]$$

$$= kL^{-1}[(kt/\alpha) \sinh(c/\alpha) \cosh(c/\alpha)]$$

$$= (k^2t^2/2!\alpha) \sinh (c/\alpha) \cosh(c/\alpha)$$

$$u_3 = kL^{-1} A_2 = kL^{-1}[(u_2/\alpha) \cosh(u_0/\alpha) + (1/2)(u_1^2/\alpha^2) \sinh(u_0/\alpha)]$$

$$= kL^{-1}[(k^2t^2/2)(1/\alpha)^2 \sinh(c/\alpha) \cosh^2(c/\alpha)$$

$$+ (1/2)(k^2t^2)(1/\alpha)^2 \sinh^3(c/\alpha)]$$

$$= (k^3t^3/3!)(1/\alpha)^2 \sinh(c/\alpha)[\sinh^2(c/\alpha) + \cosh^2(c/\alpha)]$$

$$u_4 = (k^4t^4/4!)(1/\alpha)^3[\sinh(c/\alpha)\cosh(c/\alpha)][5\sinh^2(c/\alpha) + \cosh^2(c/\alpha)]$$

.
.
.

thus, the correct solution is the sum of the u_n above while the linearized solution is

$$u = c\,[1 + kt/\alpha + (kt/\alpha)^2/2! + (kt/\alpha)^3/3! + \cdots]$$

If we assume $\alpha = c = k = 1$, we have the results $u = e^t$ in the linear case which can be compared with the nonlinear solution.

Comparison of Linear and Nonlinear Solutions

t	linear solution	nonlinear solution
0	1	1
.1	1.105170918	1.127402502
.2	1.221402758	1.278624737
.3	1.349858808	1.462380247
.4	1.491824698	1.693614375
.5	1.648721271	2.001468676
.6	1.8221188	2.456234584
.7	2.013752707	3.325545159
.75	2.117000017	4.512775469
.76	2.13827622	5.121285511
.77	2.159766254	6.939848656
.7719	2.163873711	10.9022661
.771936	2.163951611	14.69149181
.77193683	2.163953407	20.34929139
.7719368329	2.163953414	28.3241683
.7719368330	2.163953414	∞

The error in the linear solution at t is given by

t	% error
0	0
.1	1.97
.2	4.48
.3	7.69
.4	11.92
.5	17.62
.6	25.82
.7	39.45
.75	53.09
.76	58.25
.77	68.88
.7719	80.15
.771936	85.27
.77193683	89.37
.7719368329	92.36
.7719368330	100.

i.e., the linear solution converges for all t; the actual (nonlinear) solution has a singular point at $t = (\alpha/k)\ln \operatorname{ctnh}(c/2\alpha) = .7719368330$ if $\alpha = c = k = 1$ which would not be suspected from the linear solution.

To summarize, the solution of $Ly + Ny = x(t)$ is obtained by writing $y = \sum_{n=0}^{\infty} y_n$ and $Ny = \sum_{n=0}^{\infty} A_n$, where the A_n polynomials represent the specific nonlinear term N and by identifying y_0 as discussed in (1983; 1986). If $Ny = y$, the A_n degenerate simply to

y. If we linearize Ny and use the first one or two terms of an expansion, we can find the linearized substitute approximation of the A_n.

References

Adomian, G., *Nonlinear Stochastic Operator Equations*, Academic Press, 1986.

Adomian, G., *Stochastic Systems*, Academic Press, 1986.

Bellman, R. E. and G. Adomian, *Partial Differential Equations - New Methods for Their Treatment and Applications*, Reidel, 1985.

Bellomo, N. and R. Riganti, *Nonlinear Stochastic Systems in Physics and Mechanics*, World Scientific Pulbishing, 1987.

Casti, J. L., *Nonlinear System Theory*, Academic Press, 1985a.

Feynman, R. P., *The Feynman Lectures in Physics*, Addison, Wesley, 1963-65.

CHAPTER 3

Research on Initial and Boundary Conditions for Differential and Partial Differential Equations

In order to have a complete and unique mathematical description of a problem, conditions must also be specified which the required solution must satisfy. The reason, of course, is that differential and partial differential equations can have infinitely many solutions. The added conditions ensure a unique characterization of the physical problem. They can be in the form of initial or boundary conditions.

In initial condition problems, depending on the order of the differential operator, we need to specify the initial value at $t = 0$ or $t = t_0$ and also of a number of derivatives, i.e., n initial values of an nth-order ordinary differential operator.

In boundary problems, we must specify the function at two distinct points a, b. Satisfying the boundary conditions fixes the constants in the n-term approximation ϕ_n. The procedure can be clarified by a simple example:

$$d^2u/dx^2 - 40xu = 2$$

$$u(-1) = u(1) = 0$$

Let $L = d^2/dx^2$ and write

$$Lu = 2 + 40xu$$

$$u = A + Bx + L^{-1}(2) + L^{-1}(40xu)$$

Let $u_0 = A + Bx + L^{-1}(2) = A + Bx + x^2$ and $u = \sum_{n=0}^{\infty} u_n$. We carry the constants A, B along in computing the following components u_1, $u_2, \dots$. These are given by:

$$u_{n+1} = L^{-1}40xu_n$$

for $n \geq 0$. Thus

$$u_1 = L^{-1}40xu_0 = 40\, L^{-1}\, x(A + Bx + x^2)$$

We can continue in the same manner to determine the u_i as far as we like. The sum $u = \sum_{n=0}^{\infty} u_n$ is the true solution and some n-term approximation $\phi_n = \sum_{i=0}^{n-1} u_i$ will be sufficient - usually for low n. The limit of ϕ_n as $n \to \infty$ is obviously u. The constants A, B are determined by letting ϕ_n, as our approximation for u, satisfy the boundary conditions. Not only are results obtained easily, but the procedure works as well if we have nonlinear terms which are written as sums of the appropriate A_n polynomials. Finally, the procedure is easily extended to $\nabla^2 u = f + ku$ or even $\nabla^2 u + Nu = f + ku$ where Nu is an analytic term.

Now consider the problem on R^3. Let $L_x = \partial^2/\partial x^2$, $L_y = \partial^2/\partial y^2$, $L_z = \partial^2/\partial z^2$ and rewrite the equation as

$$[L_x + L_y + L_z]u = f(x,y,z) + k(x,y,z)u$$

Solve for each linear operator term in turn, thus

$$L_x u = f + ku - (L_y + L_z)u$$

$$L_y u = f + ku - (L_z + L_x)u$$

$$L_z u = f + ku - (L_x + L_y)u$$

Operate for each with the appropriate inverse, then

$$u = \phi_x + L_x^{-1} f - L_x^{-1} ku - L_x^{-1} (L_y + L_z)u$$

$$u = \phi_y + L_y^{-1} f - L_y^{-1} ku - L_y^{-1} (L_z + L_x)u$$

$$u = \phi_z + L_z^{-1} f - L_z^{-1} ku - L_z^{-1} (L_x + L_y)u$$

where ϕ_x, ϕ_y, ϕ_z are the homogeneous solutions. Add and divide by three to write a single equation in the form

$$u = u_0 + Ku$$

where

$$u_0 = (1/3)\{\phi_x + \phi_y + \phi_z + (L_x^{-1} + L_y^{-1} + L_z^{-1})f(x)\}$$

$$K = (1/3)\{(L_x^{-1} + L_y^{-1} + L_z^{-1})k(x) - L_x^{-1}(L_y + L_z)$$

$$+ L_y^{-1}(L_z + L_x) + L_z^{-1}(L_x + L_y)\}$$

With the assumption $u = \sum_{n=0}^{\infty} u_n$, we have

$$u_{n+1} = Ku_n$$

for $n \geq 0$ which determines the components after u_0. The operators L_x^{-1}, L_y^{-1}, L_z^{-1} are double integrations leading to two constants for each. These are determined by forcing ϕ_n to satisfy the given conditions.

We can have both initial and boundary conditions in a partial differential equation. Thus, for the equation

$$u_{tt} = a^2 u_{xx}$$

we could specify initial conditions (initial displacement and initial velocity, for example) given by

$$u(x,0) = \phi(x)$$

$$u_t(x,0) = \psi(x)$$

and boundary conditions at the ends a, b of an interval, e.g.,

$$u(a,t) = \alpha(t)$$

$$u(b,t) = \beta(t)$$

Other possibilities could be the specification of the motion of an endpoint with $u(a,t) = f(t)$ or the prescription of a force at an endpoint with $u_x(b,t) = g(t)$. Boundary conditions can be nonlinear also, e.g, $u_x(b,t) = f(u(b,t)) = Nu(b,t)$ where N is a nonlinear operator. The treatment of complex nonlinear, coupled, and stochastic cases is discussed in (1986) and will not be discussed again in this volume except in the context of particular applications as necessary.

Discussion of the initial/boundary conditions is sometimes minimized in physics literature, concentrating on the differential or partial differential equation of interest without stating the required supplementary conditions which make the solution unique.

Consider, for example, the heat equation in three space dimensions and time. We have:

$$c\rho\partial u/\partial t = \sum_{i=1}^{3}(\partial/\partial x_i)(k\partial u/\partial x_i)$$

where k is the coefficient of thermal conductivity, ρ is the density, and c is the specific heat at x_1, x_2, x_3. The form $u_t = \nabla^2 u$ can be obtained by the transformation $t' = kt/c\rho$. Such an equation clearly has many solutions which cannot be distinguished because of the differentiations.

To identify a particular solution from the entire set of solutions, we need supplementary conditions similar to initial/boundary conditions for ordinary differential equations. Usually, we have some stated boundary conditions, i.e., conditions specified on the boundaries of the space described by the coordinates x_1, x_2, x_3 for which the partial differential equation is applicable, and initial conditions at a particular instant $t = t_0$ or $t = 0$.

In the above problem, the temperature distribution is uniquely determined for $t > 0$ if we know the initial temperature distribution at $t = 0$ and the temperature distribution on the boundaries. If the region is unbounded, then a bounded solution is uniquely determined by the initial conditions alone.

In the one-dimensional case - considering heat flow in a finite rod of length $[0, \ell]$ - we require conditions specified at $t = 0$ and $x = 0$ and at $x = \ell$. If the far end at $x = \ell$ recedes to infinity, the physical requirement that u be zero as $x \to \infty$ means that the third condition essentially vanishes. We require only that $u \to 0$ as $x \to \infty$, i.e., our solution does not become unbounded.

Consider the Korteweg-deVries (KdV) equation:

$$u_t - 6uu_x + u_{xxx} = 0$$

In our standard notation this is

$$L_t u - 6uL_x u + L_{xxx} u = 0$$

Solution by decomposition gives:

$$u = u_0 + 6[L_t^{-1} + L_{xxx}^{-1}](uL_xu) - [L_t^{-1} L_{xxx} + L_{xxx}^{-1} L_t]u$$

where u_0 requires four supplementary conditions to be specified, after which all components are readily determined in terms of the now known u_0 as in references(1983; 1986). Physicists specify $u(x,0)$ by e.g.:

$$u(x,0) = -(1/2)a^2\text{sech}^2[(1/2)a(x - x_0)]$$

as the shape of a wave packet or soliton intended to retain its shape for all t and starting from x_0. If we specify that this shape is to be retained for all t, this is equivalent to specifying $u(0,t)$, our second condition. Also, $u(x,t)$ must have the form $u(x - ct)$ or

$$u(x,t) = -(1/2)a^2\text{sech}^2[(1/2)a(x - x_0 - ct)]$$

It is easy to see that $c = a^2$, i.e., the soliton moves to the right with velocity a^2. Of course we have not solved a partial differential equation at all with the above considerations. We have said simply that given $u(x,0) = f(x)$, then $u(x,t) = f(x - ct)$. Why write the equation at all then if only handwaving is to be done! Why is one condition *only* given in physics literature? Because one is specified; the second, or $u(0,t)$, was implicit; the third and the fourth vanish because we consider an unbounded region and the solution must be bounded. Consider Burger's equation:

$$u_t + uu_x = \upsilon u_{xx}$$

or, in our notation,

$$L_tu + uL_xu = \upsilon L_{xx}u$$

We require three conditions. We notice that in the solution by decomposition the number of conditions are apparent from the procedure for solution.

If we solve the linear diffusion equation $L_t\theta = \upsilon L_{xx}\theta$, find $\theta(x,t)$, then transform by

$$u = -2(\upsilon/\theta)(\partial\theta/\partial x)$$

we can get $u(x,t)$ for the Burger's equation. Examining the solution of the linear diffusion equation by decomposition, we see again *three* conditions are required in the u_0 term: $\theta(0,t)$, $\theta(x,0)$, and the third must vanish if $x \to \infty$. If we are given $\theta(x,0)$, $\theta(0,t)$, and $\theta_x(0,t)$, then the first term of the decomposition series is

$$\theta_0 = (1/2)[\theta(x,0) + \theta(0,t) + x\theta_x(0,t)]$$

Then the terms $\theta_1, \theta_2, \ldots$ are determined immediately from

$$(1/2)\upsilon^{-1} L_{xx}^{-1} \sum_{n=0}^{\infty} \theta_n$$

We propose to consider the KdV equation, Burger's equation, the sine-Gordon equation, the generalized heat equation, Schrödinger and nonlinear Schrödinger equation, stochastic Schrödinger equations, Navier-Stokes equation and others now with more detailed investigation.

Example:

In the example at the beginning of the chapter, we considered the equation $d^2u/dx^2 - 40\,xu = 2$ given the conditions $u(-1) = u(1) = 0$. Letting $L = d^2/dx^2$, we wrote

$$L\,u = 2 + 40\,xu$$

$$u = c_1 + c_2x + L^{-1}(2) + L^{-1}(40xu)$$

Let $u_0 = c_1 + c_2x + L^{-1}(2) = c_1 + c_2x + x^2$ and $u = \sum_{n=0}^{\infty} u_n$. The following components of u are given by

$$u_{n+1} = L^{-1}\,40xu_n$$

for $n \geq 0$, thus

$$u_1 = L^{-1}40xu_0 = (20/3)c_1x^3 + (10/3)c_2x^4 + 2x^5$$

Proceeding in the same way,

$$u_2 = (80/9)c_1x^6 + (200/63)c_2x^7 + (10/7)x^8$$

We can continue in the same manner to determine the u_i as far as we like. The sum $u = \sum_{n=0}^{\infty} u_n$ is the true solution, and some n-term approximation $\phi_n = \sum_{i=0}^{\infty} u_i$ will be sufficient - usually for low n. The constants are evaluated by using ϕ_n for u in satisfying the specified conditions.

So far, we have the three-term approximation ϕ_3 given by

$$\phi_3 = c_1 + c_2x + x^2 + (20/3)c_1x^3 + (10/3)c_2x^4 + 2x^5$$

$$+ (80/9)c_1x^6 + (200/63)c_2x^7 + (10/7)x^8$$

If we wish to impose the boundary conditions at this stage,

$$\phi_3(1) = c_1(1 + 20/3 + 80/9) + c_2(1 + 10/3 + 200/63)$$

$$+ (1 + 2 + 10/7) = 0$$

$$\phi_3(-1) = c_1(1 - 20/3 + 80/9) + c_2(- 1 + 10/3 - 200/63)$$

$$+ (1 - 2 + 10/7) = 0$$

or

$$\begin{bmatrix} 149/9 & 473/63 \\ 29/9 - & 53/63 \end{bmatrix} \begin{bmatrix} c_1 \\ c_2 \end{bmatrix} = \begin{bmatrix} -31/7 \\ - 3/7 \end{bmatrix}$$

so that

$$\begin{bmatrix} c_1 \\ c_2 \end{bmatrix} = \begin{bmatrix} -13779/75649 \\ -2034/10807 \end{bmatrix}$$

The limit of ϕ_n as $n \to \infty$ is obviously u. The results are given in Table I showing values of the n-term approximant by the decomposition method for n = 12.

Table 1

x	ϕ_{12}
-1.0	0.0
-0.8	0.254206
-0.6	0.296396
-0.2	-0.025886
0	-0.119393
0.2	-0.135649
0.6	-0.083321
0.8	-0.050944
1.0	0.0

If the analytical expression for ϕ_{12} with the evaluated constants is substituted into the left side of the differential equation, it should yield 2 exactly. Our result is 2.000000 providing seven digit accuracy. The procedure works as well in the multidimensional case, or even if we have nonlinear terms which are then written as sums of the appropriate A_n. Finally, the procedure is also easily extended to $\nabla^2 u = f(x,y,z) + k(x,y,z)u$ or even $\nabla^2 u + Nu = f(x,y,z) + k(x,y,z)u$ where Nu is an analytic term.

Nonlinear, stochastic, and coupled boundary conditions have been discussed in Adomian (1986). The approximate solution ϕ_n is substituted with the boundary conditions. Examples appear in the above reference and in a paper by Adomian and Rach (see references).

References

Adomian, G., *Nonlinear Stochastic Operator Equations*, Academic Press, New York, 1986.

Adomian, G. and R. Rach, "Coupled Differential Equations and Coupled Boundary Conditions," *J. Math. Anal. and Appl.*, 112, no. 1, 1985.

Arfken, G., *Mathematical Methods for Physicists*, Academic Press, 1971.

Bellman, R. and G. Adomian, *Partial Differential Equations - New Methods for Their Treatment and Application*, Reidel, 1985.

Koshlyakov, N., M. Smirnov, and G. Gliner, *Differential Equations of Mathematical Physics*, Interscience, 1964.

Tychonov, A. and A. Samarski, *Partial Differential Equations of Mathematical Physics*, Holden-Day, 1964.

Wylie, C., *Advanced Engineering Mathematics*, McGraw-Hill, 1975.

PART II

Applications to the Equations of Physics

Introduction: The following chapters will explore methodology for solution of nonlinear evolution equations such as the Korteweg-deVries equation, nonlinear Schrodinger equation, sine-Gordon equation, and other important nonlinear equations of physics. It is natural to inquire whether the commonly used procedure of linearization leads to the loss of physically significant features of the actual solution. Since the proposed decomposition procedure is not perturbative, it seems clear that in the propagation of solitons, for example, which depends on the nonlinear effect, the advantages are apparent. The examples will deal heavily with nonlinearity. In cases where stochasticity is involved, the procedure leads to stochastic series. In this case, the n-term approximate $\phi_n = \sum_{i=0}^{n-1} u_n$, where $u = \sum_{i=0}^{\infty} u_n$ is the desired solution, will involve stochastic processes, and desired statistics are then obtained from ϕ_n.

CHAPTER 4

The Burger's Equation

The theory of turbulence is a fascinating area of mathematical physics. A useful technique for new approaches to this difficult subject is the analysis of model equations which exhibit the essential characteristics of the more recalcitrant and realistic equations. Of such model equations, perhaps the best-known is Burger's equation. Application of the decomposition method to Burger's equation is doubly useful. It demonstrates that smallness assumptions are unnecessary, and, since Burger's equation is solvable by transformation of variables (exact linearization), it serves to validate the use of the decomposition for studies of turbulence. The equation is given as

$$u_t + uu_x = \upsilon u_{xx} \qquad x \geq 0, \; t \geq 0 \tag{4.1}$$

assuming that the necessary initial-boundary conditions are specified. Generally, the initial conditions $u(x,0)$ will be specified as a given function $g(x)$ for $x \geq 0$, but we require two more conditions for $t \geq 0$. Write (4.1) in the form $L_t u + uu_x = \upsilon L_{xx} u$ where $L_t = \partial/\partial t$, $L_{xx} = \partial^2/\partial x^2$.

Proceeding with the decomposition, we solve for the linear terms $L_t u$ and $L_{xx} u$ in turn, thus:

$$\begin{aligned} L_t u &= \upsilon L_{xx} u - uu_x \\ \upsilon L_{xx} u &= L_t u + uu_x \end{aligned} \tag{4.2}$$

Operating on the first equation of (4.2) with L_t^{-1} and on the second with L_{xx}^{-1}, we have

$$\begin{aligned} L_t^{-1} L_t u &= \upsilon L_t^{-1} L_{xx} u - L_t^{-1} uu_x \\ L_{xx}^{-1} L_{xx} u &= \upsilon^{-1} L_{xx}^{-1} L_t u + \upsilon^{-1} L_{xx}^{-1} uu_x \end{aligned} \tag{4.3}$$

L_t represents the integral operator (and can conveniently be assumed, though not necessary, to be the definite integral from 0 to t); hence

$$L_t^{-1} L_t u = u(x,t) - \phi(x)$$

where the initial condition $\phi(x) = u(x,0)$ is known. L_{xx}^{-1} is a two-fold integral operator. We will define it as a two-fold indefinite integration over x. Then

$$L_{xx}^{-1} L_{xx} u = u(x,t) - \alpha(t) - \beta(t)x$$

where α, β can be evaluated from the given conditions. Thus we have

$$u(x,t) = \phi(x) + \upsilon L_t^{-1} L_{xx} u - L_t^{-1} uu_x$$
$$u(x,t) = \alpha(t) + \beta(t)x + \upsilon^{-1} L_{xx}^{-1} L_t u + \upsilon^{-1} L_{xx}^{-1} u_x \tag{4.4}$$

We let $u = \sum_{n=0}^{\infty} u_n$ where the components are to be determined, and we identify u_0 in the first equation of (4.4) as $\phi(x)$ and u_0 in the second equation as $\alpha(t) + \beta(t)x$. Suppose, for example, we were given $u(0,t) = f(t)$ and $u(1,t) = g(t)$. Then we have $u_0 = \phi(x) = u(x,0)$ in the first equation of (4.4), and $u_0 = f(t) + (g(t) - f(t))x$ in the second equation for one-term approximations of the individual equations. (The one-term approximation of the actual solution would be half the sum of these two values for u_0, but we consider this later.) We write $\Phi_n = \sum_{i=0}^{n-1} u_i$ as the approximation to $u = \lim \Phi_n$ as $n \to \infty$ as the *solution* of each operator equation. Thus we have

$$u = u_0 + \upsilon L_t^{-1} L_{xx} u - L_t^{-1} uu_x$$
$$u = u_0 + \upsilon^{-1} L_{xx}^{-1} L_t u + \upsilon^{-1} L_{xx}^{-1} uu_x$$

replacing (4.4) where $u_0 = \phi(x)$ in the first equation and $u_0 = \alpha + \beta x$ in the second equation can be evaluated from specified condtions. Whether the problem or a similar one is defined on the real line or in a finite interval with boundary conditions, controls the evaluation of the u_0 term. Now letting $u = \sum_{n=0}^{\infty} u_n$ on both sides, and replacing the nonlinear term uu_x by $\sum_{n=0}^{\infty} A_n(u_0, u_1, \ldots, u_n)$, we have

$$u = u_0 + \upsilon L_t^{-1} L_{xx} \sum_{n=0}^{\infty} u_n - L_t^{-1} \sum_{n=0}^{\infty} A_n$$

$$u = u_0 + \upsilon^{-1} L_{xx}^{-1} L_t \sum_{n=0}^{\infty} u_n + \upsilon^{-1} L_{xx}^{-1} \sum_{n=0}^{\infty} A_n$$

Now we can identify

$$u_1 = \upsilon L_t^{-1} L_{xx} u_0 - L_t^{-1} A_0(u_0)$$

$$u_1 = \upsilon^{-1} L_{xx}^{-1} L_t u_0 + \upsilon^{-1} L_{xx}^{-1} A_0(u_0)$$

Since the A_n can be generated to represent the nonlinear term in a rapidly convergent series for the solution, we see each of the u_1 can also be calculated again yielding constants of integration to be evaluated so that the given conditions remain unchanged. Then $\Phi_2 = u_0 + u_1$ becomes a two-term approximation for each operator equation, and the approximate actual u is obtained by adding the Φ_n approximations and dividing by the number of operator equations, in this case, two. For boundary problems given conditions on an x interval and a y interval, each operator equation is handled with one set of conditions. In initial condition problems we can add the individual equations for u immediately, divide by the number of equations, and apply decomposition directly. In this case from (4.4) we can write

$$u = u_0 + (1/2)\{\upsilon L_t^{-1} L_{xx} \sum_{n=0}^{\infty} u_n - L_t^{-1} \sum_{n=0}^{\infty} A_n\{uu_x\}$$

$$+ \upsilon^{-1} L_{xx}^{-1} L_t \sum_{n=0}^{\infty} u_n + \upsilon^{-1} L_{xx}^{-1} \sum_{n=0}^{\infty} A_n\{uu_x\}\}$$

We can continue in this manner with $u_2, u_3, \ldots$. Thus,

$$u_{n+1} = \upsilon L_t^{-1} L_{xx} u_n - L_t^{-1} A_n$$

$$u_{n+1} = \upsilon^{-1} L_{xx}^{-1} L_t u_n + \upsilon^{-1} L_{xx}^{-1} A_n$$

for $n \geq 0$. Thus all components of u can be generated when the A_n are known for the specific nonlinear term, in this case uu_x, because A_n depends only on components to u_n. We list a few for convenience here.

Using primes to denote differentiations with respect to x,

$$A_0 = u_0\dot{u}_0$$

$$A_1 = u_0\dot{u}_1 + u_1\dot{u}_0$$

$$A_2 = u_2\dot{u}_0 + u_1\dot{u}_1 + u_0\dot{u}_2$$

$$A_3 = u_3\dot{u}_0 + u_2\dot{u}_1 + u_1\dot{u}_2 + u_0\dot{u}_3$$

etc. which can be written $A_n = u_n\dot{u}_0 + u_{n-1}\dot{u}_1 + \cdots + u_1\dot{u}_{n-1} + u_0\dot{u}_n$. Thus the u_n for $n = 0, 1, 2 \ldots$ are determined. Since the series obtained has been shown to be rapidly convergent in earlier work, we can expect that a finite set of terms $\phi_n = \sum_{i=0}^{n-1} u_n$ (generally with very small n) will serve as an excellent approximation to u. In the limit as n approaches infinity, $\phi_n = \sum_{i=0}^{n-1} u_i = u$.

The correct approach is always to solve for the linear terms, thus as in the above problem, we solved for $L_{xx}u$ and for $L_t u$, then inverted to get the basic equations, leaving the nonlinear term on the right expressed in the A_n polynomials.

Note that the Burger's equation $L_t u + uu_x = \upsilon L_{xx}u$ becomes the linear diffusion equation if the nonlinear term uu_x vanishes, i.e.,

$$L_t\theta = \upsilon L_{xx}\theta$$

which we can solve and then obtain u with the transformation

$$u = -(2\upsilon/\theta)(\partial\theta/\partial x)$$

If we solve the equation for $L_t\,\theta$ by decomposition,

$$\theta = \theta(x,0) + \upsilon L_t^{-1} L_{xx}\theta$$

Now solving for $L_{xx}\theta$

$$L_{xx}\theta = \upsilon^{-1} L_t\theta$$

Suppose the given conditions are $\theta(0,t)$ and $\theta_x(0,t)$

$$\theta = \theta(0,t) + x\theta_x(0,t) + \upsilon^{-1} L_{xx}^{-1}\theta$$

Thus

$$\theta = (1/2)[\theta(x,0) + \theta(0,t) + x\theta_x(0,t)] + (1/2)\upsilon^{-1} L_{xx}^{-1}\theta$$

or

$$\theta = \theta_0 + (1/2)\upsilon^{-1} L_{xx}^{-1}\theta + \upsilon L_t^{-1} L_{xx}\theta$$

where

$$\theta_0 = (1/2)[\theta(x,0) + \theta(0,t) + x\theta_x(0,t)]$$

and we can find all other components by substituting $\theta = \sum_{n=0}^{\infty} \theta_n$.

$$\theta_1 = (1/2)\ \upsilon^{-1} L_{xx}^{-1}\ \theta_0$$

$$\theta_2 = (1/2)\ \upsilon^{-1} L_{xx}^{-1}\ \theta_1$$

.

.

.

If boundary conditions are given, the second and third terms in θ_0 are written $\alpha + \beta x$ so α, β can be evaluated. However, our interest is in demonstrating the need for three conditions as we would expect by inspection of the diffusion equation. Thus, even in the linear equation, three conditions are needed. Since u can be obtained by transformation, Burger's equation needs three conditions. Similarly, in the KdV equation, which we will consider shortly, we require four conditions, not just the initial condition, often the only condition discussed.

References

Adomian, G., *Nonlinear Stochastic Operator Equations* , Academic Press, New York, 1986.

Bigi, D. and R. Riganti, "Solutions of Nonlinear Boundary-Value Problems by the Decomposition Method," *Proc. 11th IMACS World Congress on System Simulation and Scientific Computation, 4,* 1985.

Rach, R. "A Convenient Computational Form for the Adomian Polynomials," *J. Math. Anal. and Applic.* 102, 2, 1984, 415-419.

CHAPTER 5

Heat Flow and Diffusion

Partial differential equations of second order and parabolic type occur in problems involving heat flow and diffusion.

5.1 One-Dimensional Case

If we consider a single space dimension, the temperature distribution $u(x,t)$ is described by

$$(\partial/\partial x)(k\ \partial u/\partial x) + F(x,t) = c\rho\ (\partial u/\partial t) \tag{5.1.1}$$

where F represents a heat source at x at time t. If k, c, ρ are assumed constants (e.g., k is not a function of u), we have a linear problem and (5.1.1) can be given in the form

$$u_t = a^2 u_{xx} + f(x,t) \tag{5.1.2}$$

where $a^2 = k/c\rho$ and $f(x,t) = (1/c\rho)F(x,t)$ or if the source vanishes, simply

$$u_t = a^2 u_{xx} \tag{5.1.3}$$

often referred to as the heat conduction equation. We can consider, for example, a homogeneous rod of length ℓ, thermally insulated on lateral surfaces and sufficiently thin so that at any time t, the temperature $u(x,t)$ is the same at all points of the cross-section at x.

If the rod is inhomogeneous so that $k = k(x)$ and heat exchange takes place with the surrounding medium, we have an equation of the form

$$u_t - a^2 u_{xx} + \alpha u = f(x,t)$$

with $\alpha = h/c\rho$, h a heat-exchange coefficient, and

$$f(x,t) = \alpha\ T(x,t) + g(x,t)/c\rho$$

with T the temperature of the medium and $g(x,t)$ the density of the heat sources.

In the case of three space dimensions and time, the heat flow is described by $u(x,y,z,t)$ and the equation

$$c\rho u_t = \text{div}\,(k \text{ grad } u) + F \tag{5.1.4}$$

or

$$c\rho(\partial u/\partial t) = \partial/\partial x(k\ \partial u/\partial x) + \partial/\partial y(k\ \partial u/\partial y) + \partial/\partial z(k\ \partial u\partial z) + F$$

or

$$c\rho u_t = \sum_{i=1}^{3} \partial/\partial x_i(k\ \partial u/\partial x_i) + F$$

where k, c, ρ are functions of x_1, x_2, x_3. More generally we may have k dependent on u and space coordinates, e.g., in the case of large temperature fluctuation.

If the material is homogeneous, we can write

$$u_t = a^2(u_{xx} + u_{yy} + u_{zz}) + F/c\rho \tag{5.1.5}$$

where $a^2 = k/c\rho$, or,

$$u_t = a^2\nabla^2 u + f \qquad\qquad f = F/c\rho \tag{5.1.6}$$

The *diffusion equation* is analogous to the heat conduction equation. For one space dimension and time, we can write

$$\partial/\partial x(D\ \partial u/\partial x) = c(\partial u/\partial t) \tag{5.1.7}$$

or $(Du_x)_x = cu_t$. If the diffusion coefficient D is constant, we have

$$u_t = a^2 u_{xx}$$

in the same form as before with $a^2 = D/c$.

To complete the specification of the problem, we require the auxiliary initial and/or boundary conditions as well. Thus in the heat equation, we can prescribe the temperature at the ends $x = 0$ and $x = \ell$. We can specify $u(0,t) = g(t)$ where $g(t)$ is known in the time interval of interest. Or, we can specify the heat flow at one end, e.g., $\partial u(0,t)/\partial x = h(t)$. Or, we can have more complicated conditions such as

$$\partial u(\ell,t)/\partial x + \lambda u(\ell, t) = T(t)$$

These restrictions, of course, apply to all partial differential equations. Such equations have many solutions. In order to identify a particular solution from the entire possible set of solutions, we need supplementary conditions similar to specification of initial conditions in an initial-value problem described by an ordinary differential equation.

The usual situation is that we will have some boundary conditions specified on the boundaries of the space described by the coordinates x_1, x_2, x_3 or x, y, z, for which the partial differential equation is applicable and, also, initial conditions at a particular instant $t = 0$ or $t = t_0$.

If we know the temperature distribution at $t = 0$ and the temperature at the boundaries, the temperature distribution will be unequally determined for $t > 0$. If the time interval of interest is very long, the influence of the initial condition vanishes and the distribution will be determined by the boundary conditions. For $0 \leq x \leq \ell$, and $t \geq 0$, we specify $u(0,t) = f(t)$ and $u(\ell, t) = g(t)$.

If the region is unbounded, then a solution is uniquely determined by the initial conditions alone. For the bounded case, we might have conditions such as u at $t = 0$ and u at $x = 0$ and $x = a$. However, if $a \to \infty$, we retain the conditions at $t = 0$ and $x = 0$, but now $u(a) \to 0$ as $a \to \infty$. Thus, if $u = \Upsilon_1 + \Upsilon_2 x$, Υ_1, Υ_2, can be evaluated for conditions such as $u(0) = 0$ and $u(1) = 0$. If instead of $x = 1$, we have $x \to \infty$, then we can only say $u \to 0$ at $x \to \infty$ so $\Upsilon_2 = 0$.

It is clear that the influence of the initial condition vanishes for large t and the distribution is determined by the boundary conditions alone. If x lies in the semi-infinite interval $[0,\infty]$ and $t \geq 0$, we will need only $u(0,t) = f(t)$. For x in the finite interval $[0,]$ and $t \geq 0$, we can specify $u(0,t) = f(t)$ and $u(\ell, t) = g(t)$.

Suppose we consider a condition such as:

$$k\,(\partial u/\partial t)(0,t) = \sigma u^4(0,t) - T^4(0,t)$$

which occurs in an example given by Tychonov and Samarskii for heat radiation at $x = 0$ according to the Stefan-Boltzmann law in a medium with temperature T. Such conditions and methods of theoretical treatment are discussed at length in (1964) and will be taken up again here only to the extent of particular applications. Returning to problems of heat flow, we observe that the equation

$$u_t = a^2 u_{xx} + bu_x + cu$$

can be put in the form

$$v_t = a^2 v_{xx}$$

by substituting

$$u = e^{\mu x + \lambda t} v$$

$$\mu = -b/2a^2$$

$$\lambda = c - (b^2/4a^2)$$

The usual mathematical questions arise, of course: i) do solutions exist? ii) do they depend continuously on the given conditions (so that a small change in an initial or boundary condition will cause only a small change in the solution?) In the case of the heat equation, these questions have been answered affirmatively in (1964).

5.2 Two-Dimensional Case

Consider now the parabolic equation $u_{xx} + u_{yy} = ku_t$ where $k = k(x,y)$. Write this as

$$L_{xx}u + L_{yy}u = k(x,y)L_t u$$

We write three equations

$$L_{xx}u = kL_t u - L_{yy}u$$

$$L_{yy}u = kL_t u - L_{xx}u$$

$$L_t u = k^{-1}[L_{xx} + L_{yy}]u$$

Operate on the first with L_{xx}^{-1}, the second L_{yy}^{-1}, and the third with L_t^{-1} to get

$$u = u(0,y,t) + \Upsilon_1 x + L_{xx}^{-1} k L_t u - L_{xx}^{-1} L_{yy} u$$

$$u = u(x,0,t) + \Upsilon_2 y + L_{yy}^{-1} k L_t u - L_{yy}^{-1} L_{xx} u$$

$$u = u(x,y,0) + L_t^{-1} k^{-1} [L_{xx} + L_{yy}] u$$

adding and dividing by three

$$u = (1/3)\{u(0,y,t) + u(x,0,t) + u(x,y,0)\}$$

$$+ (1/3)\{\Upsilon_1 x + \Upsilon_2 y\}$$

$$+ (1/3)\{[L_{xx}^{-1} + L_{yy}^{-1}] k L_t u$$

$$- [L_{xx}^{-1} L_{yy} + L_{yy}^{-1} L_{xx}] u$$

$$+ L_t^{-1} k^{-1} [L_{xx} + L_{yy}] u\}$$

Now u is replaced by $\sum_{n=0}^{\infty} u_n$ with

$$u_0 = (1/3)\{u(0,y,t) + u(x,0,t)$$

$$+ u(x,y,0) + \Upsilon_1 x + \Upsilon_2 y\}$$

and the boundary conditions at $x = a$ and $y = b$ determine Υ_1, Υ_2. These vanish if a and b recede to infinity. Each u_n after u_0 is now determined by:

$$u_n = (1/3)\{[L_{xx}^{-1} + L_{yy}^{-1}] k L_t u_{n-1}$$

$$- [L_{xx}^{-1} L_{yy} + L_{yy}^{-1} L_{xx}] u_{n-1}$$

$$+ L_t^{-1} k^{-1} [L_{xx} + L_{yy}] u_{n-1}$$

L_{xx} involves two differentiations with respect to x. L_{yy} involves two differentiations with respect to y. L_t involves a single differentiation with respect to t. The inverses are the corresponding integrations.

A more general form with a heat source is

$$\partial^2 u/\partial x^2 + \partial^2 u/\partial y^2 - k(x,y)\, \partial u/\partial t = g$$

for $0 < t < \infty$ and $x,y \in \Omega$, a bounded rectangular region with boundary Γ with temperature distribution specified on the edge $x \in [0,a]$, and on $y \in [0,b]$, $k = k(x,y)$, given $u(x,y,0)|_\Omega = 0$ and $u(x,y,t) = f(x,y,t)$ for $x,y \in \Gamma$.

Suppose now we let k be constant and T be constant on the given surfaces. Then

$$u_0 = T - (Tx/a + Ty/b)/3$$

If a, b -> ∞, then $u_0 = T$ and

$$u_1 = 1/3[L_{xx}^{-1} + L_{yy}^{-1}]u_0 - 1/3[L_{xx}^{-1}L_{yy} + L_{yy}^{-1}L_{xx}]u_0$$
$$+ 1/3\, L_t^{-1} k^{-1} [L_{xx} + L_{yy}]u_0 = (T/6)(x^2 + y^2)$$

5.3 Three-Dimensional Case

Let us consider the more general form

$$\sum_{i=1}^{3} (\partial/\partial x_i)(k\, \partial u/\partial x_i) - c\rho\, \partial u/\partial t = g$$

where k is the coefficient of thermal conductivity, ρ is the density, and c is the specific heat at x_1, x_2, x_3. We will use x, y, z to avoid confusion with the subscripts for the decomposition. Let $L = \partial/\partial x$, $L_y = \partial/\partial y$, $L_z = \partial/\partial z$, $L_t = \partial/\partial t$. Now we have

$$L_x k L_x u + L_y k\, L_y u + L_z k\, L_z u - c\rho L_t u = g$$

Write

$$L_x k L_x u = g + c\rho L_y u - L_y k L_y u - L_z k L_z u$$

$$L_y k L_y u = g + c\rho L_t u - L_x k L_x u - L_z k L_z u$$

$$L_z k L_z u = g + c\rho L_t u - L_x k L_x u - L_y k L_y u$$

$$c\rho L_t u = - g + L_x k L_x u + L_y k L_y u + L_z k L_z u$$

The first equation becomes

$$k L_x u = u(0,y,z,t) + L_x^{-1} g + L_x^{-1} c\rho L_y u$$
$$- L_x^{-1} L_y k L_y u - L_x^{-1} L_z k L_z u$$

$$L_x u = k^{-1} [u(0,y,z,t) + L_x^{-1} g]$$
$$+ k^{-1} L_x^{-1} c\rho L_y u - k^{-1} L_x^{-1} L_y k L_y u$$
$$- k^{-1} L_x^{-1} L_z k L_z u$$

$$u = u(0,y,z,t) + L_x^{-1} \{k^{-1} [u(0,y,z,t) + L_y^{-1} g]\}$$
$$+ L_x^{-1} k^{-1} L_x^{-1} c\rho L_t u - L_x^{-1} k^{-1} L_x^{-1} L_y k L_y u$$
$$- L_x^{-1} k^{-1} L_x^{-1} L_z k L_z u$$

We similarly derive the other three equations. Thus

$$k L_y u = u(x,0,z,t) + L_y^{-1} g + L_y^{-1} c\rho L_t u - L_y^{-1} L_x k L_x u - L_y^{-1} L_z k L_z u$$

$$L_y u = k^{-1} [u(x,0,z,t) + L_y^{-1} g] + k^{-1} L_y^{-1} c\rho L_t u - k^{-1} L_y^{-1} L_x k L_x u$$
$$- k^{-1} L_y^{-1} L_z k L_z u$$

$$u = u(x,0,z,t) + L_y^{-1} k^{-1} [u(x,0,z,t) + L_y^{-1} g]$$
$$+ L_y^{-1} k^{-1} L_y^{-1} c\rho L_t u - L_y^{-1} k^{-1} L_y^{-1} L_x k L_x u$$

The third equation becomes

$$kL_z u = u(x,y,0,t) + L_z^{-1} g + L_z^{-1} c\rho L_t u$$

$$- L_z^{-1} L_x kL_x u - L_z^{-1} L_y kL_y u$$

$$L_z u = k^{-1} u(x,y,0,t) + k^{-1} L_z^{-1} g + k^{-1} L_z^{-1} c\rho L_t u$$

$$- k^{-1} L_z^{-1} L_x kL_x u - k^{-1} L_z^{-1} L_y kL_y u$$

$$u = u(x,y,0,t) + L_z^{-1} k^{-1} u(x,y,0,t) + L_z^{-1} k^{-1} L_z^{-1} g$$

$$+ L_z^{-1} k^{-1} L_z^{-1} c\rho L_t u - L_z^{-1} k^{-1} L_z^{-1} L_x kL_x u$$

$$- L_z^{-1} k^{-1} L_z^{-1} L_y kL_y u$$

The fourth equation is

$$L_t u = - g/c\rho + (1/c\rho)(L_x kL_x u) + (1/c\rho)(L_y kL_y u)$$

$$+ 1/c\rho(L_z kL_z u)$$

$$u = u(x,y,z,0) - L_t^{-1} (g/c\rho) + L_t^{-1} (1/c\rho)L_x kL_x u$$

$$+ L_t^{-1} (1/c\rho)(L_y kL_y u)$$

Adding, dividing by four, and replacing u by $\sum_{n=0}^{\infty} u_n$, and defining u_0 to include all the auxiliary conditions and terms involving g, we have

$$u_0 = (1/4)\{u(0,y,z,t) + u(x,0,z,t) + u(x,y,0,t)$$

$$+ u(x,y,z,0) + L_x^{-1} k^{-1} u(0,y,z,t)$$

$$+ L_y^{-1} k^{-1} u(x,0,z,t) + L_z^{-1} k^{-1} u(x,y,0,t)$$

$$- L_t^{-1} (g/c\rho) + L_z^{-1} k^{-1} L_z^{-1} g + L_y^{-1} k^{-1} L_y^{-1} g + L_x^{-1} k^{-1} L_x^{-1} g\}$$

Now all remaining terms can be found from

$$
\begin{aligned}
u_{n+1} = (1/4)\{ & L_x^{-1} k^{-1} L_x^{-1} c\rho L_t - L_x^{-1} k^{-1} L_x^{-1} k L_y \\
& - L_x^{-1} k^{-1} L_x^{-1} L_z k L_z + L_y^{-1} k^{-1} L_y^{-1} c\rho L_t \\
& - L_y^{-1} k^{-1} L_y^{-1} L_x k L_x + L_z^{-1} k^{-1} L_z^{-1} c\rho L_t \\
& - L_z^{-1} k^{-1} L_z^{-1} L_x k L_x - L_z^{-1} k^{-1} L_z L_y k L_y \\
& + L_t^{-1} (1/c\rho) L_x k L_x + L_t^{-1} (1/c\rho) L_y k L_y \} u_n
\end{aligned}
$$

for $n \geq 0$.

If the scale is finite, we might be given conditions at the edges farthest from the origin and flow conditions involving derivatives at the near edges. Let's assume k, c, ρ are constants. Then we have the form:

$$a^2 \nabla^2 u - u_t = f$$

Now define $L_x = \partial^2/\partial x^2$, $L_y = \partial^2/\partial y^2$, $L_z = \partial^2/\partial z^2$, $L_t = \partial/\partial t$. Then

$$a^2 [L_x + L_y + L_z] u - L_t u = f$$

Write the four equations

$$a^2 L_x u = f + L_t u - a^2 L_y u - a^2 L_z u$$

$$a^2 L_y u = f + L_t u - a^2 L_x u - a^2 L_z u$$

$$a^2 L_z u = f + L_t u - a^2 L_x u - a^2 L_y u$$

$$L_t u = - f + a^2 L_x u + a^2 L_y u + a^2 L_z u$$

or

$$L_x u = a^{-2} f + a^{-2} L_t u - L_y u - L_z u$$

$$L_y u = a^{-2}f + a^{-2}L_t u - L_x u - L_z u$$

$$L_z u = a^{-2}f + a^{-2}L_t u - L_x u - L_y u$$

$$L_t u = -f + a^2 L_x u + a^2 L_y u + a^2 L_z u$$

For simplicity let us allow $a = 1$ here. Then

$$L_x u = f + L_t u - L_y u - L_z u$$

$$L_y u = f + L_t u - L_x u - L_z u$$

$$L_z u = f + L_t u - L_x u - L_y u$$

$$L_t u = -f + L_x u - L_y u - L_z u$$

Operating with the inverse operators (two-fold integration for x, y, z and single for t),

$$u(x,y,z,t) = A + Bx + L_x^{-1} f + [L_x^{-1} L_t - L_x^{-1} L_y - L_x^{-1} L_z]u$$

$$u(x,y,z,t) = C + Dy + L_y^{-1} f + [L_y^{-1} L_t - L_y^{-1} L_x - L_y^{-1} L_z]u$$

$$u(x,y,z,t) = E + Fz + L_z^{-1} f + [L_z^{-1} L_t - L_z^{-1} L_x - L_z^{-1} L_y]u$$

$$u(x,y,z,t) = G - L_t^{-1} f + [L_t^{-1} L_x + L_t^{-1} L_y + L_t^{-1} L_z]u$$

$A, B, \ldots, G$ are chosen to satisfy boundary conditions. Now we have

$$u_0 = (1/4)\{A + Bx + C + Dy + E + Fz + G - [L_x^{-1} + L_y^{-1} + L_z^{-1} - L_t^{-1}]f\}$$

$$u_{n+1} = (1/4)\{L_x^{-1} L_t - L_x^{-1} L_y - L_x^{-1} L_z + L_y^{-1} L_t - L_y^{-1} L_x - L_y^{-1} L_z + L_z^{-1} L_t - L_z^{-1} L_x - L_z^{-1} L_y + L_t^{-1} L_x + L_t^{-1} L_y + L_t^{-1} L_z\}\, u_n$$

for $n \geq 0$.

5.4 Some Examples

Consider the equation

$$\nabla^2 u - \partial u/\partial t = g(x,y,z,t) \tag{5.4.1}$$

Define the linear operators $L_x = \partial^2/\partial x^2$, $L_y = \partial^2/\partial y^2$, $L_z = \partial^2/\partial z^2$, $L_t = \partial/\partial t$, and write (5.4.1) as

$$[L_x + L_y + L_z]u - L_t u = g \tag{5.4.2}$$

We solve in turn for $L_x u$, $L_y u$, $L_z u$, $L_t u$. Thus

$$\begin{aligned}
L_x u &= g - L_y u - L_z u + L_t u \\
L_y u &= g - L_z u - L_x u + L_t u \\
L_z u &= g - L_x u - L_y u + L_t u \\
L_t u &= -g + L_x u + L_y u + L_z u
\end{aligned} \tag{5.4.3}$$

for which

$$\begin{aligned}
u &= \Upsilon_1 + \Upsilon_2 x + L_x^{-1} g - L_x^{-1} L_y u - L_x^{-1} L_z u + L_x^{-1} L_t u \\
u &= \Upsilon_3 + \Upsilon_4 y + L_y^{-1} g - L_y^{-1} L_z u - L_y^{-1} L_x u + L_y^{-1} L_t u \\
u &= \Upsilon_5 + \Upsilon_6 z + L_z^{-1} g - L_z^{-1} L_x u - L_z^{-1} L_y u - L_z^{-1} L_t u \\
u &= \Upsilon_7 - L_t^{-1} g + L_t^{-1} L_x u + L_t^{-1} L_y u + L_t^{-1} L_z u
\end{aligned} \tag{5.4.4}$$

where the Υ's are not presumed to be numerical constants. They depend on the conditions specified on the solution. Adding, we have

$$\begin{aligned}
u = u_0 + (1/4)\{L_t^{-1}[L_x + L_y + L_z] - L_x^{-1}[L_y + L_z + L_t] \\
- L_y^{-1}[L_z + L_x + L_t] - L_z^{-1}[L_x + L_y + L_t]\}u
\end{aligned} \tag{5.4.5}$$

with

$$u_0 = (1/4)\{\Upsilon_1 + \Upsilon_2 x + \Upsilon_3 + \Upsilon_4 y + \Upsilon_5 + \Upsilon_6 z + \Upsilon_7 + [L_x^{-1} + L_y^{-1} + L_z^{-1} - L_t^{-1}]\, g\}$$

The homogeneous solutions and therefore the Υ's must be evaluated from the given conditions on u. Once u_0 is completely specified, u can be completely determined as discussed in (1985;1964) by assuming $u = \sum_{n=0}^{\infty} u_n$. Then

$$u_{n+1} = (1/4)\{L_t^{-1}[L_x + L_y + L_z] - L_x^{-1}[L_y + L_z + L_t] - L_y^{-1}[L_z + L_x + L_t] - L_z^{-1}[L_x + L_y + L_t]\}u_n \tag{5.4.6}$$

for $n \geq 0$. The expression

$$\phi_n = \sum_{i=0}^{n-1} u_i \tag{5.4.7}$$

is the approximation to u. A natural question is the number of terms n required to represent u to a desired accuracy. This question has been considered in (1985; 1964) and will not be dealt with here except to say that easily computable and accurate solutions have been obtained with small n for differential and partial differential equations, even when nonlinear terms or stochastic processes are included. Numerical computations clearly demonstrate the convergence to desired accuracy.

Let us take the homogeneous heat equation $u_{xx} - u_t = 0$ as a specific mathematical example with given conditions as an illustration of the procedure. We assume $u(0,t) = t$, $u(x,0) = x^2/2$, and $\partial u/\partial x|_{x=0} = 0$; we now have

$$u = u(0,t) + L_x^{-1} L_t u$$

$$u = u(x,0) + L_t^{-1} L_x u$$

and

$$u = (1/2)[u(0,t) + u(x,0)] + (1/2)[L_x^{-1} L_t + L_t^{-1} L_x] \sum_{n=0}^{\infty} u_n$$

Now

$$u_0 = (1/2)[t + (x^2/2)]$$

$$u_1 = (1/2)[L_x^{-1} L_t + L_t^{-1} L_x]u_0$$

$$u_2 = (1/2)[L_x^{-1} L_t + L_t^{-1} L_x]u_1$$

.
.
.

$$u_{n+1} = (1/2)[L_x^{-1} L_t + L_t^{-1} L_x]u_n \qquad n \geq 0$$

This specific case with the given conditions is chosen for each evaluation and

$$[L_x^{-1} L_t + L_t^{-1} L_x]\,[u(0,t) + u(x,0)]$$

$$= L_x^{-1} L_t u(0,t) + L_x^{-1} L_t u(x,0) + L_t^{-1} L_x u(0,t) + L_t^{-1} L_x u(x,0)$$

$$= L_x^{-1} L_t(t) + L_x^{-1} L_t(x^2/2) + L_t^{-1} L_x(t) + L_t^{-1} L_x(x^2/2)$$

$$= (x^2/2) + t = u(x,0) + u(0,t)$$

Thus *for this case* , $[L_x^{-1} L_t + L_t^{-1} L_x]$ is an identity operator, a special case making the calculation particularly simple. In the general case one has to compute the effects of these operators repeatedly, but it is still simple. Now

$$u_0 = (1/2)[t + (x^2/2)]$$

$$u_1 = (1/2)[L_x^{-1} L_t + L_t^{-1} L_x]u_0 = (1/4)[t + (x^2/2)]$$

$$u_2 = (1/2)[L_x^{-1} L_t + L_t^{-1} L_x]u_1 = (1/8)[t + (x^2/2)]$$

.
.
.

$$u_n = (1/2^{n+1})[t + (x^2/2)]$$

Hence,

$$u = [t + (x^2/2)] \sum_{n=0}^{\infty} (1/2^{n+1})$$

is the solution. Denoting an n-term approximation by ϕ_n,

$$\phi_1 = 0.5(t + x^2/2) = u_0$$

$$\phi_2 = 0.75(t + x^2/2) = u_0 + u_1$$

$$\phi_3 = 0.875(t + x^2/2) = u_0 + u_1 + u_2$$

$$\vdots$$

$$u = \lim_{n\to\infty} \phi_n = u_1 + u_1 + \ldots + u_{n-1} = (1)(t + x^2/2)$$

which satisfies the problem. Note that six terms yield the solution to better than 98% and with 10 terms, the approximation is within 99.9% of the correct value.

Addition of stochasticity in g, or as a coefficient α of the u_t term, or of nonlinear terms is easily handled by generalizations discussed in (1964).

Let us consider a more physical example. For one-dimensional flow parallel to the x-axis, the equation is $u_t = ku_{xx}$ where k is the thermal diffusivity. Taking $k = 1$, we have $u_t = u_{xx}$ if heat is transferred only by conduction, and no source is present. Boundary conditions describing the thermal conditions on the surface of the solid and the initial temperature distribution are necessary along with the heat equation to determine the temperature distribution $u(x,t)$. If, for example, we have a three-dimensional solid with an insulated end at $x = 0$, the condition on u is $u_x(0,y,z,t) = 0$ because the heat flow across the surface would be proportional to $u_x(0,y,z,t)$. Suppose we have a bar whose temperature distribution is described by $u(x,t)$. Let the bar be initially at temperature $f(x) = \sin x$ and assume the ends at $x = 0$ and $x = 1$ are kept at temperature zero. We have therefore the conditions $u(x,0) = \sin x$, and $u(0,t) = u(1,t) = 0$.

If we proceed blindly, we get

$$u_0 = (1/2) \sin x$$

$$u_1 = (1/4)\{L_x^{-1} L_t + L_t^{-1} L_x\} \sin x$$

.
.
.

If we continue in this way, we should get either no convergence or extremely slow convergence. The problem is that in our two equations - one derived using the appropriate inverse on the equation for $L_t u$ and the other derived analogously from the equation for $L_x u$, both do not contribute with these conditions. The one equation is

$$u = u(x,0) + L_t^{-1} L_x u$$

$$= \sin x + L_t^{-1} L_x u$$

so that

$$u_0 = \sin x$$

$$u_{n+1} = L_t^{-1} L_x u_n$$

which immediately gives us $u = e^{-t} \sin x$ as the correct solution. The other equation is

$$u = A + Bx + L_x^{-1} L_t u$$

but satisfying the boundary conditions with $u_0 = A + Bx$, we have u_0 f= 0, thus the equations does not contribute and is not used. The correct solution $u = e^{-t} \sin x$ is easily verified.

It is interesting also to consider steady-state temperature in a bar with ends at $x = 0$ and $x = 1$, maintained at $u = 0$ and $u = T$ respectively. Then $u(0) = 0$ and $u(1) = T$. Now $u = u_0 = Ax + B$ where $B = 0$ and $A = T$ so that $u = Tx$ is the solution.

5.5 Heat Conduction in an Inhomogeneous Rod

Consider a rod of length ℓ lying along the x axis with its left end at the origin. Suppose its density is a function of x, its cross-section is A, its mass per unit length is ρ, specific heat is c, and conductivity is k_1. Let $u(x,t)$ represent temperature at a point x at time t and let $F(x,t)$ be the amount of heat per unit cross-section per unit time passing x to the right. The cylinder of base area A between the points x and $x + \delta$ has heat supplied to it at the rate $c\rho\delta A u_t = AF(x,t) - AF(x + \delta,t) = - AF_x\delta$ where higher powers of δ are dropped. Since $F = - k_1u_x$ in heat conduction, $F_x = - k_1u_{xx}$. Now

$$(c\rho/k_1)u_t = u_{xx}$$

or letting $k = c\rho/k_1$, we have the heat equation

$$u_{xx} = k(x)u_t \tag{5.5.1}$$

Suppose we have initial boundary conditions $u(x,0) = g(x)$ for $0 < x < 1$ and $u(0,t) = h_1(t)$, $u(1,t) = h_2(t)$ for $t > 0$. Then

$$u_0 = (1/2)[g(t) + h_1(t) + x(h_2(t) - h_1(t))] \tag{5.5.2}$$

from which the following terms can be computed.

Since we previously treated k as a constant, some modification is needed. Thus

$$L_xu = k(x)L_tu$$

or

$$u = A + Bx + L_x^{-1} k(x)L_tu$$

and

$$k(x)L_tu = L_xu$$

$$L_tu = k^{-1}(x)L_xu$$

$$u = C + L_t^{-1} k^{-1}(x)L_xu$$

$$u = C + L_t^{-1} k^{-1}(x)L_xu$$

so that

$$u = (1/2)[A + Bx + C] + (1/2)[L_x^{-1} k(x) L_t + L_t^{-1} k^{-1}(x)L_x]u$$

$$u_0 = (1/2)[A + Bx + C] \qquad (5.5.3)$$

$$u_{n+1} = (1/2)[L_x^{-1} k(x) L_t + L_t^{-1} k^{-1} (x)L_x]u_n$$

Now we can consider some specific examples. Let $h_1 = h_2 = 0$ and $g(x) = \sin \pi x$. For these specific b.c., we again have a case where one equation does not contribute. Thus $u_0 = \sin \pi x$ and not $(1/2) \sin \pi x$. For the case $k(x) = k^2$, a constant,

$$u_0 = \sin \pi x$$

$$u_1 = L_t^{-1} (1/k^2)L_x u_0 = -(\pi^2 t/k^2) \sin \pi x$$

$$u_2 = L_t^{-1} (1/k^2)L_x u_1$$

$$\vdots$$

$$u = [1 - (\pi^2 t/k^2) + (\pi^4 t^2/2k^4) - \ldots]\sin \pi x$$

$$u = e^{-\pi^2 t/k^2} \sin \pi x$$

Thus we satisfy the b. c. for the u_0 from each equation. If they are non-vanishing, we can add and divide by the number of equations. We can proceed with determination of components for each equation separately to some n-term approximation then add and divide to get an approximation to the total solution.

To consider more general cases we can use (5.5.3) with a given $k(x)$, e.g., $k(x) = 1 + x$, and compute a series for $u(x,t)$. The method is easily generalized now to parabolic equations such as

$$u_{xx} + u_{yy} = ku_t$$

$$\nabla^2 u - ku_t = g$$

where $k = k(x,y)$ or $k(x,y,z)$ and $g = g(x,y,z,t)$ or even stochastic cases.

5.6 Nonlinear Heat Conduction

Consider the equation

$$(\partial/\partial x)[k(u)\, \partial u/\partial x] = c\rho(\partial u/\partial t)$$

$$u(0,t) = f(t)$$

$$u(x,0) = g(x)$$

$$u_x(0,t) = 0$$

We suppose f, g are analytic functions, and we can let $k(u)$ be any analytic function - a linear function $k_0 + k_1 u$, or $\alpha + \beta u + \Upsilon u^2$, or $\sinh u$. Suppose we consider the first which is sufficient to demonstrate a nonlinear equation. We have

$$L_x k(u) L_x u = c\rho L_t u$$

or

$$L_x k_0 L_x u + L_x k_1 u L_x u = c\rho L_t u \tag{5.6.1}$$

The procedure is straightforward. We have two linear terms $k_0 L_x L_x u$ and $c\rho L_t u$ and the nonlinear term $k_1 L_x(u L_x u)$. In the nonlinear term, we replace $uL_x u = uu_x$ by $\sum_{n=0}^{\infty} A_n$ where the A_n are generated for $Nu = uu_x$. We then solve for the two linear terms in turn using the inverse operators L_x^{-1} and L_t^{-1}. We define u_0 as usual and assume the decomposition $u = \sum_{n=0}^{\infty} u_n$. The components $u_1, u_2, \ldots$ now are determinable.

Solving (5.6.1) for the linear term $L_x k_0 L_x u$ and substituting α for $c\rho$,

$$k_0\ L_x\ L_x u = \alpha L_t u - k_1\ L_x u\ L_x u$$

$$L_{xx} u = k_0^{-1} \alpha\ L_t u - k_0^{-1}\ k_1\ L_x u\ L_x u$$

$$u = A + Bx + k_0^{-1} \alpha\ L_{xx}^{-1}\ L_t u - k_0^{-1}\ k_1\ L_{xx}^{-1}\ L_x u\ L_x u$$

Now $B = 0$ and $A = u(0,t)$ hence

$$u = u(0,t) + k_0^{-1} \alpha L_{xx}^{-1} L_t u - k_0^{-1} k_1 L_{xx}^{-1} L_x u L_x u \qquad (5.6.2)$$

Now, solving for the remaining linear term [on the right side of (5.6.1)]

$$L_t u = \alpha^{-1} L_x k_0 L_x u + \alpha^{-1} L_x k_1 u L_x u$$

$$u = u(x,0) + \alpha^{-1} k_0 L_t^{-1} L_{xx} u + \alpha^{-1} k_1 L_t^{-1} L_x u L_x u \qquad (5.6.3)$$

Adding (5.6.2), (5.6.3) and dividing by two:

$$u_0 = (1/2)\{u(0,t) + u(x,0)\}$$

$$u_1 = (1/2)\{\alpha^{-1} k_0 L_t^{-1} L_{xx} u_0 + k_0^{-1} \alpha L_{xx}^{-1} L_t u_0$$
$$+ (\alpha^{-1} k_1 L_t^{-1} - k_0^{-1} k_1 L_{xx}^{-1})(u u_{xx} u_x u_x)\}$$

The nonlinear quantity $(u u_{xx} u_x u_x)$ is written in terms of the A_n for all of the u_n terms; in this term it is A_0.

Now

$$u = (1/2)\{(\alpha^{-1} k_0 L_t^{-1} L_{xx} + k_0^{-1} \alpha L_{xx}^{-1} L_t) u_0$$
$$+ (\alpha^{-1} k_1 L_t^{-1} - k_0^{-1} k_1 L_{xx}^{-1}) A_0\}$$

$$u_{n+1} = (1/2)\{(\alpha^{-1} k_0 L_t^{-1} L_{xx} + k_0^{-1} \alpha L_{xx}^{-1} L_t) u_n$$
$$+ (\alpha^{-1} k_1 L_t^{-1} - k_0^{-1} k_1 L_{xx}^{-1}) A_n\}$$

5.7 Heat Conduction Equation with Discontinuous Coefficients

Suppose the heat conducting material is not homogeneous and the coefficients in the equation are discontinuous at one or even several points. Divide the total interval $[0,\ell]$ at the points ε_i of discontinuity into subintervals. If the temperature and the heat flow are continuous at the points ε_i, we have

$$u(\varepsilon_i - 0,t) = u(\varepsilon_i + 0,t)$$

$$k(\varepsilon_i - 0)\ \partial u/\partial x(\varepsilon_i - 0,t) = k(\varepsilon_i - 0)\ \partial u/\partial x(\varepsilon_i + 0,t)$$

5.8 Nonlinear Boundary Conditions

If heat is being radiated at the $x = 0$ cross section of the conducting material with the temperature $T(t)$, we have the nonlinear boundary condition (1964).

$$k\ \partial u/\partial x(0,t) = \sigma[u^4(0,t) - T^4(0,t)]$$

Such boundary conditions are treated in (1986).

5.9 Comparisons

The power and convenience of the decomposition .method is shown by the following examples illustrating some comparisons with established procedures.

Consider the homogeneous heat conduction equation

$$u_t = a^2 u_{xx}$$

$$u(x,0) = \phi(x)$$

$$u(0,t) = u(\ell,t) = 0$$

and apply the well-known separation of variables technique assuming $\phi(x)$ is continuous, bounded, possesses piecewise continuous derivatives and satisfies $\phi(0) = \phi(\ell) = 0$. (Then $u(x,t)$ will be continuous for $t \geq 0$.) Let $u(x,t) = X(x)T(t)$. Then we have

$$(1/a^2)(T'/T) = (X''/X) = -\lambda = \text{constant}$$

which leads to the equations

$$X'' + \lambda X = 0 \qquad X(0) = X(\ell) = 0$$

$$\lambda_n = (\pi n/\ell)^2 \qquad n = 1, 2, 3, \ldots$$

$$X_n(x) = \sin(\pi n x/\ell)^2$$

$$T_n(t) = C_n e^{-a \lambda_n t}$$

where the C_n are constants. Thus

$$u(x,t) = \sum_{n=0}^{\infty} X_n(x) T_n(t) = \sum_{n=0}^{\infty} C_n e^{-a^2 \lambda_n t} \sin(\pi n/\ell)x$$

satisfies the homogeneous boundary conditions. To satisfy the initial conditions

$$\phi(x) = u(x,0) = \sum_{n=1}^{\infty} C_n \sin(\pi n x/\ell)$$

which is the Fourier sine series so that the C_n are Fourier coefficients

$$C_n = \phi_n = 2/\ell \int_0^{\ell} \phi(\varepsilon)\sin(\pi n \varepsilon/\ell)d\varepsilon$$

[It can be shown that the series converges and is appropriately (term-wise twice) differentiable with respect to x, and satisfies the differential equation in $0 < x < \ell$, $t > 0$.]

The solution can also be written as

$$u(x,t) = \int_0^{\ell} G(x,\varepsilon,t)\phi(\varepsilon)d\varepsilon$$

when the Green's function G is given by

$$G = (2/\ell) \sum_{n=1}^{\infty} \exp\{-(\pi n/\ell)^2 a^2 t\} \sin(\pi n x/\ell) \sin(\pi n \varepsilon/\ell)$$

which can be seen by looking at $u(x,t)$, substituting values of C_n. Since the series for C_n converges uniformly for $t > 0$ with respect to ε, the summation and the integration can be interchanged.

If we consider

$$u_t = a^2 u_{xx} + f(x,t)$$

$$u(x,0) = u(0,t) = u(\ell,t) = 0$$

the solution is

$$u(x,t) = \int_0^t \int_0^\ell G(x,\varepsilon,t-f)(\varepsilon,\tau)d\varepsilon d\tau$$

$$G = (2/\ell) \sum_{n=1}^{\infty} e^{-(\pi n/\ell)^2 a^2 (t-\tau)} \sin(\pi n x/\ell) \sin(\pi n \varepsilon/\ell)$$

We can, if we wish, apply decomposition to the separated equation, but we emphasize that there is no need for the separation anymore since we can do the partial differential equation directly. It is interesting, however, to consider it because the equation

$$y'' + \omega^2 y = 0 \qquad\qquad y(0) = y(\ell) = 0$$

appears to be a case where decomposition does not work. The problem is that not only is there no forcing function but the homogeneous solution vanishes. Thus y_0 term is zero, so all components vanish. Such a case is discussed in (1986) Since there is no y_0, we will add a linear term $\beta x + \Upsilon$. Since $y(0) = 0$, $\Upsilon = 0$ and βx is sufficient . If we satisfy $y(\ell) = 0$, we have nothing so let us defer that. Then

$$y = \beta x - \omega^2 L^{-1} \sum_{n=0}^{\infty} y_n$$

$$y_0 = \beta x$$

$$y_1 = -\omega^2 L^{-1} y_0 = -\omega^2 L^{-1} \beta x = -\omega^2 \beta x^3/3!$$

$$y_2 = -\omega^2 L^{-1} y_2 = \omega^4 \beta x^5/5!$$

.
.
.

Thus $y = \beta x - \omega^2\beta x^3/3! + \omega^4\beta x^5/5! - \ldots$. The β must be ω if the differential equation is to be satisfied so that the series is clearly $y = \sin \omega x$. Satisfying $y(\ell) = 0$ means $\omega = n\pi x/\ell$ and $y = \sin(n\pi x/\ell)$.This is an interesting area for further work for students.

5.10 Uncoupled Equations with Coupled Conditions

Such a situation arises in a melting problem. Let $x \geq 0$ be the material to be melted. At $t = 0$ the $x = 0$ surface is at a temperature $T >> T_m$, the melting temperature. As the material melts, the transition surface between solid and liquid penetrates into the solid material for $t > 0$. Denote this moving boundary or transition surface by $x = \varepsilon$. $x = \varepsilon$ moves from x_1 to $x_2 = x_1 + \Delta\varepsilon$ during an interval $[t, t + \Delta t]$ in which an amount $\rho\Delta\varepsilon$ melts. Let k_1 be the thermal conductivity of the liquid state and k_2 the thermal conductivity of the solid state. Then

$$\{k_1 (\partial u/\partial x)\big|_{x_1} - k_2 (\partial v/\partial x\big|_{x_2} \}\Delta t = \lambda\rho\Delta\varepsilon$$

or, as $\Delta\varepsilon \to 0$

$$k_1 L_x u\big|_{x=\varepsilon} - k_2 L_x v\big|_{x=\varepsilon} = \lambda L_t \varepsilon$$

As we see here, u and v can enter in a single boundary condition. Let $a^2 = kc\rho$ be the coefficient of temperature conductivity in the equation $u_t = a^2 u_{xx} + f(x,t)$ with different values in two regions

$$u_t = a^2 u_{xx} \quad \text{for } 0 < x < \varepsilon$$

$$v_t = a^2 v_{xx} \quad \text{for } \varepsilon < x < \infty$$

or letting $L_x u = u_{xx}$ and $L_t u = u_t$ and similarly $L_x v = v_{xx}$ and $L_t v = v_t$, we have

$$L_t u = a^2 L_x u \qquad\qquad 0 < x < \varepsilon$$

$$L_t v = a^2 L_x v \qquad \varepsilon < x < \infty$$

Exercise:

Solve the above. A useful reference is in (1986).

References

Adomian, G. , *Nonlinear Stochastic Operator Equations* , Academic Press, New York, 1986.

Bellman, R. E., and G. Adomian. *Partial Differential Equations - New Methods for Their Treatment and Applications* , Reidel, 1985.

Bellomo, N., L. deSocio, and R. Monaco, "On the Random Heat Equation: Solution by the Stochastic Adaptive Interpolation Method," *Comp. and Math. with Appl.* , 1987.

Tychonov, A. N., and A. A. Samarskii, *Partial Differential Equations of Mathematical Physics* , Holden-Day, 1964.

CHAPTER 6

Nonlinear Oscillations in Physical Systems

6.1 Oscillatory Motion

Practically all the problems of mechanics are nonlinear at the outset. Nonlinear oscillating systems are generally analyzed by approximation methods which involve some sort of linearization. These replace an actual nonlinear system with a so-called "equivalent" linear system and employ averagings which are not generally valid.

While the linearizations commonly used are adequate in some cases, they may be grossly inadequate in others since essentially new phenomena (shock waves in gas dynamics, for example) can occur in nonlinear systems which cannot occur in linear systems. A correct solution of a nonlinear system is much more significant a matter than simply getting more accuracy when we solve the nonlinear system rather than a linearized approximation. Thus, if we want to know how a physical system behaves, it is essential to retain the nonlinearity, not just solve a convenient mathematized model. Physical problems are nonlinear; linearity is a special case just as a deterministic system is a special case of a stochastic system. In a linear system, cause and effect are proportional. Such a linear relation rarely holds. The general case is nonlinear and stochastic. In such cases, it is usual to make many restrictive assumptions - often with very little attempt at physical justification. Using the decomposition method, these restrictions are unnecessary; correct solutions are obtained in the strongly nonlinear case and in the case of stochastic (large fluctuation) behavior, as well as in the cases where perturbation would be applicable or in the linear and/or deterministic limits. It should not require justification to present the correct solution to the actual physical problem. What does require continuing justification are "smallness" assumptions, linearized models, and the assumption of artifical and often physically unrealistic processes for the convenience of mathematicians and the utilization of available theorems.

Here we are concerned with the study of vibrations which is

concerned with oscillatory motion and the associated forces. Vibrations can occur in any mechanical system having mass and elasticity. Consequently, they can occur in structures and machines of all kinds. In large space structures containing men and machines, such vibrations will result in difficult and crucial control problems and also lifetime or duration considerations since vibrations can lead to eventual failure.

Oscillations can be regular and periodic, or they can be random as in an earthquake. Randomness leads to stochastic differential equations. In deterministic systems - the special case where randomness vanishes - the equations modeling the phenomena or system provide instantaneous values for any time. When random functions are involved, the instantaneous values are unpredictable, and it is necessary to resort to a statistical description. Such random functions of time, or stochastic processes, occur in problems, for example, such as pressure gusts encountered by aircraft, jet engine noise, or ground motion in earthquakes.

If we model oscillatory motion, we get equations of the form:

$$\ddot{y} + f(y,\dot{y},t) = x(t)$$

$$\mathcal{F}\, y = x(t)$$

Stochastic processes may be involved in coefficients, input terms, or initial boundary conditions, so $x(t)$ will be assumed to be generally stochastic. Input conditions - possibly stochastic - are given (statistically described if stochastic). $\mathcal{F}$ is a nonlinear stochastic operator (1983). More conveniently, we write $\mathcal{F}\, y = \mathcal{L}\, y + \mathcal{N}\, y$ where $\mathcal{L}\, y$ is a linear (stochastic) term, and $\mathcal{N}\, y$ is a nonlinear (stochastic) term, or

$$\mathcal{L}\, y + \mathcal{N}\, y = x$$

Write $\mathcal{L} = L + R + \mathcal{R}$ where L is a linear invertible deterministic operator, so L^{-1} exists, R is the remainder of the linear deterministic operator, vanishing, of course, if we can easily invert the entire linear deterministic operator, and $\mathcal{R}$ is a (linear) stochastic operator. Thus L may be $\langle \mathcal{L} \rangle$ if it is invertible, and if not, we write $L + R$. Similarly, $\mathcal{N}\, y = Ny + \mathcal{M}\, y$ since there may be both deterministic nonlinear terms Ny and stochastic (nonlinear) terms $\mathcal{M}\, y$. In general we have

$$Ly + Ry + \mathcal{R}y + Ny + \mathcal{M}y = x$$

where, in a particular equation, any number of terms from one to four may vanish in a particular problem. Still more generally Ny may actually be a function of $y, \dot{y}, \ldots$ but again, it is simply a matter of calculating the A_n; this can be done even for a composite nonlinearity, and we will simply write Ny without further ado. Although convergence of the decomposition series can reasonably be expected to be fastest when we invert the entire linear deterministic operator, computation of the integrals will, of course, be more difficult since we will not then have simple Green's functions. Thus, generally, we will let L denote the highest-order differential operator or d^2/dt^2 in the above equation.

In an oscillator, we have generally an external force or driving term $x(t)$, a restoring force $f(y)$ dependent on the displacement y, and a damping force, since energy is always dissipated in friction or resistance to motion. Usually this is dependent on velocity, and we will write it as $g(\dot{y})$.

If we have a free oscillating mass m on a spring with no damping, we can write

$$my'' + ky = 0$$

if the spring obeys Hooke's Law, i.e., assuming displacement proportional to force. Of course, no spring really behaves this way. Often the force needed for a given compression is not the same as for an extension of the same amount. Such asymmetry is represented by a quadratic force, or force proportional to y^2 rather than y. We may have a symmetric behavior but proportionality to y^3. Then the solution is not the harmonic solution which one gets for the model equation $my'' + ky = 0$ although it is still a periodic solution. The damping force $g(\dot{y})$ may be $c\dot{y}$ where c is constant, or it may be more complicated such as $g(\dot{y}, \dot{y}^2)$ so that it depends on v^2 as well as v. By usual methods, analytic solutions then become impossible.

Suppose we write $-f(y)$ for the restoring force, $-g(\dot{y})$ for the damping force, and represent the driving force with g; the resulting equation will be

$$y'' + f(y) + h(\dot{y}) = g$$

Suppose the restoring force is represented by an odd function such that $f(y) = -f(-y)$. We have this in most applications; it means simply that if we reverse the displacement, then the restoring force reverses its direction. A pendulum, for example, behaves this way. We might take the first two terms of the power series for $f(y)$ and write $f(y) = \alpha y + \beta y^3$. Then we have

$$y'' + \alpha y + \beta y^3 = g$$

If we have damping also, we have

$$y'' + c\dot{y} + \alpha y + \beta y^3 = g$$

assuming the damping force is $-c\dot{y}$. This is Duffing's equation which we will discuss, but it should be obvious now that we can consider much more general oscillatior equations with any $f(y)$.

We will begin - after a simple example of a pendulum (harmonic motion) - with the well-known anharmonic oscillator then go on to consider more general oscillators such as the Duffing oscillator and the Van der Pol oscillator. The Duffing oscillator in a random force field is modeled by $y'' + \alpha y' + \beta y + \Upsilon y^3 = x(t)$. It can be analyzed without limiting the force $x(t)$ to a white noise or restricting α, β, Υ to be deterministic. The same applies to the Van der Pol oscillator modeled by $y'' + \varepsilon y^2 y' - \varepsilon y' + y = x(t)$. These equations are in our standard form $\mathcal{F} y = x(t)$ which can be solved by the decomposition method. If the equation is linear and deterministic, we have $Fy = (L + R)y = x$. An equation that is deterministic but nonlinear is $Fy = Ly + Ny = x$. A linear stochastic equation is $\mathcal{L}\, y = x$, etc. These cases are discussed in the earlier work. Consider the pendulum problem as an example, then we proceed with the anharmonic oscillator.

6.2 Pendulum Problem

Consider the simple vertical pendulum consisting of a mass m at the end of a rod length ℓ moved through an angle θ. We obtain immediately

$$d^2\theta/dt^2 + k^2 \sin\theta = 0$$

where $k^2 = g/\ell$. Let $L = d^2/dt^2$ and $N(\theta) = k^2 \sin\theta$ to obtain our usual standard form $L\theta + N\theta = 0$ for a homogeneous deterministic differential equation. The usual treatment is to simplify this anharmonic motion by assuming $\sin\theta \approx \theta$. We can then write $d^2\theta/dt^2 + k^2\theta = 0$, the well-known harmonic oscillator problem. L is defined as before, but now there is no nonlinear term, and the remainder of the linear term is $R\theta$, i.e., the entire linear operator is decomposed into $L + R$ where R is a linear deterministic "remainder" operator. Define L^{-1} as the double definite integration from 0 to t. If we have a forcing term as well, we would have $L\theta + R\theta = x(t)$. The solution is found by writing

$$L\theta = x - R\theta$$

$$L^{-1}L\theta = \theta - \theta(0) - t\theta'(0) = L^{-1}x - L^{-1}R\theta$$

$$\theta(t) = \theta_0 - L^{-1}R\theta$$

with $\theta_0 = \theta(0) + t\dot{\theta}(0) + L^{-1}x(t)$. Now substituting $\theta = \sum_{n=0}^{\infty} \theta_n(t)$, we find

$$\theta_{n>1} = - L^{-1}R\theta_n$$

so that all components are determined.

Let us assume initial conditions are given as $\theta(0) = \Upsilon$ and $\dot{\theta}(0) = 0$. Then, since $x = 0$

$$\theta_0 = \Upsilon$$

$$\theta_1 = - L^{-1}R\theta_0 = - L^{-1}k^2\Upsilon = - \Upsilon k^2t^2/2!$$

$$\theta_2 = - L^{-1}R\theta_1 = - L^{-1}k^2(-\Upsilon k^2t^2/2!) = \Upsilon k^4t^4/4!$$

$$\theta_3 = - \Upsilon k^6t^6/6!$$

We have

$$\theta_n = \Upsilon(-1)^{n-1}(kt)^{2n-2}/(2n-2)!$$

$$= \Upsilon \sum_{n=1}^{\infty} (-1)^{n-1}(kt)^{2n-2}/(2n-2)!$$

Thus $\theta = \Upsilon \cos kt$ with $k = (g/\ell)^{1/2}$.

If we assume initial conditions $\theta(0) = 0$, $\dot{\theta}(0) = \omega$, we obtain

$$\theta_0 = \omega t = (\omega/k)kt$$

$$\theta_1 = -(\omega/k)k^3t^3/3!$$

$$\theta_2 = (\omega/k)k^5t^5/5!$$

$$\vdots$$

$$\theta_n = (\omega/k)(-1)^{n-1}(kt)^{2n-1}/(2n-1)!$$

and finally,

$$\theta(t) = (\omega/k)\{kt - (kt)^3/3! + (kt)^5/5! - \ldots\}$$

$$= (\omega/k) \sum_{n=1}^{\infty} (-1)^{n-1} (kt)^{2n-1}/(2n-1)!$$

$$\theta(t) = (\omega/k) \sin kt$$

The case $\theta(0) = \Upsilon$, and $\dot{\theta}(0) = \omega$ yields

$$\theta_0 = \Upsilon + (\omega/k)kt$$

$$\theta_1 = -L^{-1}R[\Upsilon + (\omega/k)kt]$$

$$\vdots$$

$$\theta_n = \Upsilon(-1)^{n-1} (kt)^{2n-2}/(2n-2)!$$

$$+ (\omega/k)(-1)^{n-1} (kt)^{2n-1}/(2n-1)!$$

or

$$\theta(t) = \Upsilon \cos kt + (\omega/k) \sin kt$$

the well-known general solution of the harmonic oscillator. We can generalize to the anharmonic oscillator describing the pendulum problem for large amplitude motion, the only difference being that since we now have a nonlinearity, it must be expressed in terms of the A_n polynomials.

Anharmonic Oscillator: Consider the equation for large-amplitude motion in the pendulum

$$d^2\theta/dt^2 + k^2 \sin \theta = 0$$

with $k^2 = g/\ell$. Let $L\theta = d^2\theta/dt^2$ and $N(\theta) = k^2 \sin \theta$. We let $N\theta = \sum_{n=0}^{\infty} A_n$ where the A_n are given by:

$$A_0 = \sin \theta_0$$

$$A_1 = \theta_1 \cos \theta_0$$

$$A_2 = -(\theta_1^2/2) \sin \theta_0 + \theta_2 \cos \theta_0$$

$$A_3 = -(\theta_1^3/6) \cos \theta_0 - \theta_1\theta_2 \sin \theta_0 + \theta_3 \cos \theta_0$$

$$\vdots$$

This equation is a nonlinear deterministic homogeneous second-order differential equation. We must assume appropriate initial conditions. We will choose here $\theta(0) = \Upsilon$ = constant and $\dot{\theta}(0) = 0$. We can do it just as well for other combinations of initial conditions, of course, as will be obvious. We now have

$$L\theta = -N(\theta)$$

Operating with L^{-1}, and noting that $L^{-1}u$ exists if u is measurable in some time interval, we now have

$$\theta = \theta_0 - L^{-1}N(\theta)$$

where

$$\theta_0 = \theta(0) + t\dot{\theta}(0)$$

which in our case is just Υ. The $N(\theta)$ term - we will henceforth drop the parentheses - is replaced by the $\sum_{n=0}^{\infty} A_n$. We now have

$$\theta = \Upsilon - L^{-1} \sum_{n=0}^{\infty} A_n(k^2 \sin \theta)$$

We prefer to write

$$\theta = \Upsilon - L^{-1} k^2 \sum_{n=0}^{\infty} A_n (\sin \theta)$$

i.e., we will let $N\theta = \sin \theta$ rather than $k^2 \sin \theta$. The corresponding $A_n(\sin \theta)$, or simply A_n, are (1983):

$$A_0 = \sin \theta_0$$

$$A_1 = \theta_1 \cos \theta_0$$

$$A_2 = - (\theta_1^2/2) \sin \theta_0 + \theta_2 \cos \theta_0$$

$$A_3 = - (\theta_1^3/6) \cos \theta_0 - \theta_1\theta_2 \sin \theta_0 + \theta_3 \cos \theta_0$$

$$A_4 = (\theta_1^4/24)\sin \theta_0 - (\theta_1\theta_2^2/2) - [(\theta_2^2/2) + \theta_1\theta_3]\sin \theta_0 - \theta_4 \cos \theta_0$$

$$\vdots$$

Hence,

$$\theta_0 = \Upsilon$$

$$\theta_1 = - L^{-1}A_0 = - L^{-1}k^2 \sin \theta_0 = - (\sin \Upsilon)\, k^2t^2/2!$$

$$\theta_2 = - L^{-1}A_1 = - L^{-1}k^2\, \theta_1 \cos \theta_1 = (k^4t^4/4!) \sin \Upsilon \cos \Upsilon$$

$$\theta_3 = - (k^6t^6/6!) \cdot [\sin \Upsilon \cos^2 \Upsilon - 3 \sin^3 \Upsilon]$$

etc. Finally,

$$\theta(t) = \Upsilon - [(kt)^2/2!]\sin \Upsilon + [(kt)^4/4!]\sin \Upsilon \cos \Upsilon$$

$$- [(kt)^6/6!][\sin \Upsilon \cos^2 \Upsilon - 3 \sin^3 \Upsilon]$$

$$+ [(kt)^8/8!][-33 \sin^3 \Upsilon \cos \Upsilon + \sin \Upsilon \cos^3 \Upsilon] - \ldots$$

6.3 The Duffing and Van der Pol Oscillators

In the preceding section the decomposition method was applied to harmonic and anharmonic oscillators; now we apply it to the Duffing oscillator and the Van der Pol oscillator for solutions without linearization or "smallness" assumptions. When we consider, for example, the equation for a simple pendulum, we usually approximate $\sin x$ by x to obtain the harmonic oscillator equation. Suppose we go a step further and write $\sin x = x - x^3/3!$, i.e., use the first two terms of the series for $\sin x$ assuming small x. We then get the equation $x'' + \omega^2 x + \varepsilon x^3 = 0$, which is the Duffing equation with ε as a "small" parameter. This is a perturbation method, and one seeks a solution in the form $x(t) = x_0(t) + \varepsilon x_1(t) + \ldots$. We will consider such equations without smallness assumptions.

In dealing with stochastic oscillators, we depart again from usual procedures which require some sort of approximation in order to determine the second-order response statistics. A common procedure in this connection is *statistical linearization* . This procedure simply replaces the original nonlinear equation with a so-called "equivalent" linear system. Thus, if we write an oscillator equation in the form:

$$x'' + \alpha x' + \omega_0 x + \beta f(x) = F(t)$$

where $x(t)$ is a displacement, α is a damping constant, ω_0 is a linear frequency, $\beta f(x)$ is a nonlinear restoring force, and $F(t)$ is a stationary process, the process of statistical linearization substitutes

$$x'' + \alpha x + \Upsilon^2 x = F(t)$$

where Υ^2 is determined in such a way that the mean square error due to the replacement is minimized, and the mean displacement is the same for both systems. It is customary to assume $F(t)$ is Gaussian and delta-correlated with zero-mean, or, $\langle F(t)\rangle = 0$ and $\langle F(t)F(t')\rangle = 2D\delta(t-t')$. This latter assumption is, of course, made for mathematical, not physical, reasons and is physically unrealistic. We propose none of these restrictions and will solve the actual nonlinear equation.

The *Duffing oscillator* is described by the equation:

$$y'' + \alpha y' + \beta y + \Upsilon y^3 = x(t) \tag{6.3.1}$$

in our standard $F\, y = x(t)$ form.

The *Van der Pol equation* is generally given as:

$$y'' + \varepsilon y^2 y' - \varepsilon y' + y = x(t)$$

or by

$$y'' + \varepsilon y'(y^2 - 1) + y = x(t)$$

which we rewrite again as

$$y'' + \alpha y' + \beta y + \Upsilon(d/dt)y^3 = x(t) \tag{6.3.2}$$

since $y^2 y' = (d/dt)(y^3/3)$. Thus $\alpha = -\varepsilon$, $\beta = 1$, $\Upsilon = \varepsilon/3$ relates (6.3.2) to the two previously given forms. Since (6.3.1), (6.3.2) are now in our standard form $F\, y = L\, y + N\, y = x$ form, or $Fy = Ly + Ny = x$ if no stochasticity is involved, we will consider them together. (L can always be written $L + R$ for simpler inversion.) We will consider the equations to be deterministic here; we generalize later.

The linear operator in both equations is given by $d^2/dt^2 + \alpha d/dt + \beta$ The nonlinear term Ny is a simple cubic nonlinearity Υy^3 in the case of the Duffing oscillator, and $\Upsilon(d/dt)y^3$ in the case of the Van der Pol oscillator. These terms will, of course, be expanded in our A_n polynomials generated for the specific nonlinearity.

The treatment of the linear operator offers some alternatives. We can use the entire linear operator as L which enhances speed of convergence, but the inverse and consequent integrations become more difficult. We can also use part of the above operator which could be $L = d^2/dt^2$, $L = d^2/dt^2 + \alpha d/dt$, or $L = d^2/dt + \beta$. We prefer in most cases to use $L = d^2/dt^2$, i.e., the highest order differential operator. We expect this to give the slowest convergence but much easier integrations and less actual computation time. The remainder of the linear operator will be called R, the "remainder" operator. If $L = d^2/dt^2$, $R = \alpha d/dt + \beta$. (When we consider stochasticity, we will use a script letter $\mathcal{R}$ for a random part of the operator and may have $L + R + \mathcal{R}$.)

The choice made here that $L = d^2/dt^2$ results in the simplest computation. Generally, this choice of the highest ordered derivative for L is the most desirable because the integrations are the simplest. If we invert the entire linear operator, convergence is expected to be much faster. It is interesting to examine a compromise here which can be used to advantage on occasion.

If we choose $L = d^2dt^2 + \beta$, $R = \alpha d/dt$, we gain something in convergence rate over the previous case and expect to lose something in easy computability. The interesting aspect that suggests the compromise is that we see we will get sine and cosine functions for solutions of the homogeneous equation.

For the Duffing equation, we now have

$$Ly = x - Ry - \Upsilon y^3$$

$$y = c_1\phi_1(t) + c_2\phi_2(t) + L^{-1}x - L^{-1} Ry - \Upsilon L^{-1}y^3$$

where ϕ_1, ϕ_2 satisfy $L\phi = 0$ or $d^2\phi/dt^2 + \beta\phi = 0$. Consequently,

$$\phi_1(t) = \cos\sqrt{\beta}t$$

$$\phi_2(t) = (1/\sqrt{\beta}) \sin \sqrt{\beta}t$$

Now

$$\sum_{n=0}^{\infty} y_n = c_1\phi_1(t) + c_2\phi_2(t) + L^{-1}x - L^{-1}Ry - \Upsilon L^{-1} \sum_{n=0}^{\infty} A_n$$

or

$$y_0 = c_1 \cos \sqrt{\beta}t + (c_2/\sqrt{\beta}) \sin \sqrt{\beta}t + L^{-1}x$$

$$y_{n+1} = - L^{-1}Ry_n - \Upsilon L^{-1} A_n = - \alpha L^{-1}(d/dt)y_n - \Upsilon L^{-1} A_n$$

where the A_n are the appropriate polynomials for $Ny = y^3$. These are given by:

$$A_0 = y_0^3$$

$$A_1 = 3y_0^2y_1$$

$$A_2 = 3y_0y_1^2 + 3y_0^2y_2$$

$$A_3 = y_1^3 + 6y_0y_1y_2 + 3y_0^2y_3$$

$$A_4 = 3y_1^2y_2 + 3y_0y_2^2 + 6y_0y_1y_3 + 3y_0^2y_4$$

$$A_5 = 3y_1y_2^2 + 3y_1^2y_3 + 6y_0y_2y_3 + 6y_0y_1y_4 + 3y_0^2y_5$$

$$A_6 = y_2^3 + 6y_1y_2y_3 + 3y_1^2y_4 + 3y_0y_3^2 + 6y_0y_2y_4 + 6y_0y_1y_5 + 3y_0^2y_6$$

$$\vdots$$

Since $L = d^2/dt^2 + \beta$ now, L^{-1} is no longer the simple two-fold integral, and we must determine the Green's function for this L. G will satisfy the equation $LG(t,\tau) = \delta(t - \tau)$ or

$$d^2G(t,\tau)/dt^2 + \beta G(t,\tau) = \delta(t - \tau)$$

G, of course, is determinable in a number of ways. We will again use the decomposition method itself and write

$$d^2G/dt^2 = \delta(t - \tau) - \beta G$$

so we again have a simple second-order operator to invert. Hence

$$G(t,\tau) = G(0,\tau) + tG_t(0,\tau) + L^{-1}\delta(t - \tau) - \beta L^{-1}\sum_{n=0}^{\infty} G_n$$

Thus

$$\sum_{n=0}^{\infty} G_n = G(0,\tau) + tG_t(0,\tau) + L^{-1}\delta(t - \tau) - \beta L^{-1}\sum_{n=0}^{\infty} G_n$$

$$= G_0 - \beta L^{-1}\sum_{n=0}^{\infty} G_n$$

where

$$G_0 = G(0,\tau) + tG_t(0,\tau) + tH(t - \tau)$$

and

$$G_1 = -\beta L^{-1} G_0 = -\beta\{G(0,\tau)t^2/2! + G_t(0,\tau)t^3/3!$$

$$+ (t^3/3!)H(t-\tau)\}$$

$$= -\beta G(0,\tau)[t^2/2! + t^3/3!] - \beta(t^3/3!)H(t-\tau)$$

etc., for G_2, G_3, ... An appropriate n-term approximation can now be used in the L^{-1} integrations.

The example $my'' + \omega^2 y + \alpha y^3 = 0$ occurs in the theory of nonlinear vibrating mechanical systems and in some nonlinear electrical systems which is the Duffing case above. Suppose we have the specified conditions $y = a$ at $t = 0$ and $\dot{y} = 0$ at $t = 0$. Write $L = d^2/dt^2$ and

$$Ly = -(\omega^2/m)y - (\alpha/m)y^3$$

$$y = y(0) - L^{-1}(\omega^2/m)\sum_{n=0}^{\infty} y_n - (\alpha/m)\sum_{n=0}^{\infty} A_n$$

where

$$y_0 = a$$

$$y_1 = -(\omega^2/m)L^{-1}y_0 - (\alpha/m)A_0$$

$$= -a\omega^2t^2/2m - \alpha a^3/m$$

.

.

so that $y = a[1 - \omega^2t^2/2m - \alpha a^2/m \ldots]$.

For the Duffing equation with $L = d^2/dt^2$ we have

$$y_0 = y(0) + ty'(0) + L^{-1}x(t)$$

$$y_1 = -L^{-1}\alpha(d/dt)y_0 - L^{-1}\beta y_0 - L^{-1}\Upsilon A_0$$

$$y_2 = -L^{-1}\alpha(d/dt)y_1 - L^{-1}\beta y_1 - L^{-1}\Upsilon A_1$$

$$y_3 = -L^{-1}\alpha(d/dt)y_2 - L^{-1}\beta y_2 - L^{-1}\Upsilon A_2$$

etc.

For the Van der Pol equation we have

$$y_0 = y(0) + ty'(0) + L^{-1}x(t)$$

$$y_1 = -L^{-1}\alpha(d/dt)y_0 - L^{-1}\beta y_0 - L^{-1}\Upsilon(d/dt)A_0$$

$$y_2 = -L^{-1}\alpha(d/dt)y_1 - L^{-1}\beta y_1 - L^{-1}\Upsilon(d/dt)A_1$$

Stochastic Case: We could have stochastic fluctuations in α, β, or Υ in addition, of course, to stochastic $x(t)$ or initial conditions. Thus, in general, we could write

$$\alpha = \langle\alpha\rangle + \varepsilon$$

$$\beta = \langle\beta\rangle + \eta$$

$$\Upsilon = \langle\Upsilon\rangle + \sigma$$

where ε, η, σ are zero-mean random processes. The solution can now be obtained from

$$Ly = x - \langle\alpha\rangle(d/dt)y - \langle\beta\rangle y - \langle\Upsilon\rangle \sum_{n=0}^{\infty} A_n$$

$$- \varepsilon(d/dt)y - \eta y - \sigma \sum_{n=0}^{\infty} A_n$$

where the A_n summation represents y^3 in the Duffing case and $(d/dt)y^3$ in the Van der Pol case. Thus,

$$y_0 = y(0) + ty'(0) + L^{-1}x$$

$$y_1 = -\langle\alpha\rangle(d/dt)y_0 - \langle\beta\rangle y_0 - \langle\Upsilon\rangle A_0$$

$$- \varepsilon(d/dt)y_0 - \eta y_0 - \sigma A_0$$

$$y_2 = -\langle\alpha\rangle(d/dt)y_1 - \langle\beta\rangle y_1 - \langle\Upsilon\rangle A_1$$

$$- \varepsilon(d/dt)y_1 - \eta y_1 - \sigma A_1$$

.
.
.

Then $y(t) = \sum_{n=0}^{\infty} y_n(t)$ yields a stochastic series from which statistics can now be obtained without problems of statistical separability of quantities such as $\langle R\, y \rangle$ where $R = \varepsilon(d/dt) - \eta$ which normally require closure approximations and truncations (see discussion of the hierarchy method in 1983).

Remark: When transient behavior becomes oscillatory and periodic after a certain time, we can then consider it as a boundary-value problem. Sometimes one can obtain a solution for the oscillating behavior directly starting with the transient behavior. In other cases it may have to be approached separately.

References

Adomian, G., "Decomposition Solution for Duffing and Van der Pol Oscillators," *Int. J. Math. and Math. Sci.*, 9,4, 1986, 731-732.

Adomian, G., "Nonlinear Oscillations in Physical Systems," *Math. and Computers in Simulation* , 29, 1987, 275-284.

Adomian, G. *Stochastic Systems* , Academic Press, 1983.

Adomian, G., "Vibration in Offshore Structures," Part I and Part II, *Math. Comput. Simulation* , 29, 1987.

Adomian, G. and R. Rach, "An Algorithm for Transient Dynamic Analysis," Trans. Soc. for Computer Simul. 2,4, 321-327.

Adomian, G. and R. Rach, "Anharmonic Oscillator Systems," *J. Math. Analysis and Appl.* , 91, no., 1983, 229-236.

Bonzani, I., "Analysis of Stochastic Van der Pol Oscillators using the Decomposition Method," in *Complex and Distributed Systems: Analysis, Simulation and Control* , ed. S. Tzafestas and P. Borne, Amsterdam, 1986.

Gabetta, E., "On a Class of Semilinear Stochastic Systems in Mechanics with Quadratic Type Nonlinearities," J. Math. Anal. and Applic., in press.

CHAPTER 7

The KdV Equation

In the last two decades, it has become increasingly important in many areas of physics and engineering to study nonlinear and/or stochastic systems and processes arising in diverse areas, e.g., in the theory of plasmas, hydrodynamics, and aerodynamics. Among these are nonlinear wave equations such as the Korteweg-deVries (KdV) equation, the nonlinear Schrodinger equation, the sine-Gordon equation, and many others. The assumptions in equations such as the normally-stated wave equation $\psi_{tt} = \alpha^2\psi_{xx}$ describing propagation of waves traveling with velocity α, may well be at considerable variance with the physical situation. The assumptions are:

1) The amplitude of oscillations is "small".
2) Nonlinear terms in ψ can be disregarded or linearized.
3) There is no dissipation, and there exists invariance with respect to time inversion.
4) There is no dispersion - the propagation velocity is not dependent on wavelength which is generally equivalent to the assumption of large wavelength.

Thus, the form of the equation does not depend at all on properties of the medium; just as in Maxwell's equations, one never seems to see ε dependent on E or μ dependent on H but only as unquestioned constants. If these effects are significant, each medium must be dealt with separately, so it is convenient to use one equation for a wide class of phenomena which becomes more an exercise in pure mathematics than a part of physics. Some of these questions will be discussed further in the following chapters.

Let us consider the KdV equation which arises in a number of physical problems associated with waves. It describes wave propagation (in one direction) in shallow water, or magnetohydrodynamic waves in a cold plasma, etc. It is an important example of a nonlinear dispersive system. The solution to the initial- value problem was obtained by Miura, Gardner, Kruskal, et al. Here, we seek a solution without linearization, using decomposition. The equation can be written $u_t + \alpha uu_x + \beta u_{xxx} = 0$, since scaling of the variables allows choosing α, β arbitrarily. Often α is taken as -6 and β as 1. Thus we have

$$u_t + 6uu_x + u_{xxx} = 0$$

with the nonlinear term uu_x. The term u_{xxx} is a dispersion term. This requires four conditions. The exact solution is obtained for a special class of solutions by an exact linearization of the initial-value problem given the initial condition $u(x,0) = f(x)$ with $-\infty < x < \infty$ and $f(x)$ rapidly approaching zero as $|x| \to \infty$. Write $L_t = \partial/\partial t$, $L_x = \partial/\partial x$, $L_{xxx} = \partial^3/\partial x^3$. We have two linear terms $L_t u$ and (the dispersion term) $L_{xxx}u$. Solving for the linear terms in turn

$$L_t u = 6uL_x u - L_{xxx}u \tag{7.1}$$

$$L_{xxx}u = 6uL_x u - L_t u \tag{7.2}$$

Operating on (7.1) with L_t^{-1}, we have

$$u = u(x,0) + 6L_t^{-1}(uL_x u) - L_t^{-1}L_{xxx}u \tag{7.3}$$

Operating on (7.2) with L_{xxx}^{-1},

$$u = \alpha + \beta x + \gamma x^2 + 6\,L_{xxx}^{-1}uL_x u - L_{xxx}^{-1}L_t u \tag{7.4}$$

The constants of integration are determined from specified conditions. If $x \in I = [0, \ell]$, we need the boundary conditions at $x = 0$ and $x = \ell$ in addition to the initial condition $u(x,0)$.

Now in each of the equations for u, we assume the decomposition of $u = \sum_{n=0}^{\infty} u_n$ and write $Nu = \sum_{n=0}^{\infty} A_n$ where the A_n have been defined so that each A_n for any n involves components u_0 to u_n and are generated for the specific nonlinear term $Nu = -6uL_x u$ as discussed in the referenced works. We now have

$$u = u(x,0) + L_t^{-1}\sum_{n=0}^{\infty}A_n - L_t^{-1}L_{xxx}\sum_{n=0}^{\infty}u_n \tag{7.5}$$

$$u = \alpha + \beta x + \gamma x^2 + L_{xxx}^{-1}\sum_{n=0}^{\infty}A_n - L_{xxx}^{-1}L_t\sum_{n=0}^{\infty}u_n \tag{7.6}$$

In (7.5) we identify $u_0 = u(x,0)$ and

$$\begin{aligned} u_1 &= L_t^{-1} A_0 - L_t^{-1} L_{xxx}\, u_0 \\ u_2 &= L_t^{-1} A_1 - L_t^{-1} L_{xxx}\, u_1 \\ u_3 &= L_t^{-1} A_2 - L_t^{-1} L_{xxx}\, u_2 \\ &\;\vdots \end{aligned} \tag{7.7}$$

In (7.6) we identify $u_0 = \alpha + \beta x + \gamma x^2$ and

$$\begin{aligned} u_1 &= L_{xxx}^{-1} A_0 - L_{xxx}^{-1} L_t u_0 \\ u_2 &= L_{xxx}^{-1} A_1 - L_{xxx}^{-1} L_t u_1 \\ u_3 &= L_{xxx}^{-1} A_2 - L_{xxx}^{-1} L_t u_2 \\ &\;\vdots \end{aligned} \tag{7.8}$$

The final solution u, or actually an n-term approximation ϕ_n which approaches u as $n \to \infty$, is obtained by addition of (7.5) and (7.6) and dividing by two to get a single equation which we now write with the caution that, dependent on the given conditions, we will generally evaluate (7.5) and (7.6), or equivalently (7.7) and (7.8) separately for ϕ_n for each before combining. The combined equation is:

$$\begin{aligned} u = {} & (1/2)\{u(x,0) + \alpha + \beta + \gamma x^2\} \\ & - (1/2)\{L_t^{-1} L_{xxx} + L_{xxx}^{-1} L_t\}u \\ & + (1/2)\{L_t^{-1} + L_{xxx}^{-1}\}Nu \end{aligned} \tag{7.9}$$

u is to be viewed as a decomposition of the solution into components to be determined with u_0 determined by the specified auxiliary conditions, the inverse operator, and the forcing function if one is present. Nu will be represented exactly by the infinite sum of the A_n polynomials.

$$u = u_0 - (1/2)\{L_t^{-1} L_{xxx} + L_{xxx}^{-1} L_t\}\sum_{n=0}^{\infty} u_n$$
$$+ (1/2)\{L_t^{-1} + L_{xxx}^{-1}\}\sum_{n=0}^{\infty} A_n \qquad (7.10)$$

From (7.10) all other components of u are determined since u_0 is known. Thus

$$u_1 = -(1/2)\{L_t^{-1} L_{xxx} + L_{xxx}^{-1} L_t\}u_0$$
$$+ (1/2)\{L_t^{-1} + L_{xxx}^{-1}\}A_0$$

$$u_2 = -(1/2)\{L_t^{-1} L_{xxx} + L_{xxx}^{-1} L_t\}u_1$$
$$+ (1/2)\{L_t^{-1} + L_{xxx}^{-1}\}A_1$$

$$\vdots$$

$$u_{n+1} = -(1/2)\{L_t^{-1} L_{xxx} + L_{xxx}^{-1} L_t\}u_n$$
$$+ (1/2)\{L_t^{-1} + L_{xxx}^{-1}\}A_n$$

Thus once the $A_n(u_0, u_1, \ldots u_n)$ are specified, we can calculate the approximation ϕ_n to desired n. Thus, given the initial/boundary condtions, the solution is obtained. The constants can be evaluated for each u_n or for ϕ_n all at once in each of the two equations. They converge to the constants for the complete solution u. The complete "solution" is again the sum of the two ϕ_n divided by two. The A_n are easily calculated as:

$$A_0 = -6\{u_0 L_x u_0\}$$

$$A_1 = -6\{u_0 L_x u_1 + u_1 L_x u_0\}$$

$$A_2 = -6\{u_0 L_x u_2 + u_1 L_x u_1 + u_2 L_x u_0\}$$

$$A_3 = -6\{u_0 L_x u_3 + u_1 L_x u_2 + u_2 L_x u_1 + u_3 L_x u_0\}$$

$$\vdots$$

$$A_n = -6\{u_0 L_x u_n + \cdots + u_n L_x u_0\}$$

Given the first case where boundary conditions are specified, we will evaluate the components u_n in the two equations for u separately, thus $u_0 = u(x,0)$ from the first and $u_0 = \alpha(t) + \beta(t)x + \gamma(t)x^2$. If the interval is infinite, $u_0 = u(0,t)$. In any case u_0 can be evaluated. Thus in one case, we have:

$$u_1 = L_t^{-1} A_0 - L_t^{-1} L_{xxx} u_0$$

and, in the second case, we have:

$$u_1 = L_{xxx}^{-1} A_0 - L_{xxx}^{-1} L_t u_0$$

The integration "constants" are evaluated then we calculate u_2 for each equation.

The solution, of course, is obtained as in the initial-condition case by adding the two n-term approximations and dividing by two. Thus, the solution can be investigated for various specified conditions. For some, no solution will exist. An equation for which $u_0 = 0$ is disregarded; it does not contribute to the total solution. For proper conditions, a unique solution can exist. These matters remain to be studied.

If we are given $u = 0$ at $x = 0$, and u and u_x are zero at ℓ, we have a solvable system. If $\ell \to \infty$, we need only $u(0,t)$ in addition to the initial condition $u(x,0)$. Still, another possibility is a periodic initial condition, e.g., $u(x,0) = \cos x$ and periodic boundary conditions $u(x,t) = u(x + a, t)$.

We assume the initial condition is given as

$$u(x,0) = -(1/2)a^2 \operatorname{sech}^2 [(1/2)a(x - x_0)]$$

for a soliton at $x = x_0$ with the above shape. If we assume that the shape is retained, we can say $u(x,t)$ is of the form $u(x - ct)$ or

$$u(x,t) = -(1/2)a^2 \operatorname{sech}^2 [(1/2)a(x - x_0 - a^2 t)]$$

so that the soliton moves to the right with a velocity a^2. Retention of shape as propagation takes place seems to make equation (7.3) sufficient for description specifically. Actually the assumption is equivalent to specifying $u(0,t)$, our second needed condition.

$$u(0,t) = -(1/2)a^2 \operatorname{sech}^2 [-(1/2)(x_0 + a^2 t)]$$

The remaining two terms, which we tentatively wrote as initial condition terms in (7.4), should have been written as $\beta x + \Upsilon x^2$ where the constants β and Υ must satisfy boundary conditions. If the boundary conditions are satisfied at infinity, $\beta = \Upsilon = 0$. Consequently

$$u_0 = (1/2)\{u(x,0) + u(0,t)\}$$

and the problem can be solved in a straightforward way.

What has been accomplished? Physicists sometimes write a partial differential equation with only a given initial waveform and no other stated conditions. This assumes the shape is maintained. If the solution is assumed, a differential equation is unnecessary. Complete specification involves four conditions.

In Burger's equation $L_t u + Nu = \upsilon L_{xx} u$, with $Nu = uL_x u$, three conditions are required. If the Nu is omitted, we have the linear diffusion equation

$$L_t \theta = \upsilon L_{xx} \theta$$

To find the solution $\theta(x,t)$ for initial conditions

$$\theta = \theta(x,0) + \upsilon L_t^{-1} L_{xx} \theta$$

$$\theta = \theta(0,t) + x\theta_x(0,t) + \upsilon^{-1} L_{xx}^{-1} \theta$$

Adding and dividing by two

$$\theta = (1/2)\{\theta(x,0) + \theta(0,t) + x\theta_x(0,t)\} + (1/2)\,\upsilon^{-1} L_{xx}^{-1} \theta$$

or

$$\theta_0 = (1/2)\{\theta(x,0) + \theta(0,t) + x\theta_x(0,t)\}$$

$$\theta_1 = (1/2)\,\upsilon^{-1} L_{xx}^{-1} \theta_0$$

$$\theta_2 = (1/2)\, \upsilon^{-1} L_{xx}^{-1} \theta_1$$

.
.
.

We can also solve Burger's equation using $u = -2(\upsilon/\theta)(\partial\theta/\partial x)$. Clearly, three conditions are required in Burger's equation and four in the KdV equation.

A final remark: All examples of nonlinear dispersive waves employ tedious *perturbation* calculations to derive the equation of interest such as the KdV equation. Thus, what is really needed is a new modeling with awareness that neither perturbation nor linearization or discretization, nor the neglect of stochastic fluctuations is necessary.

The KdV equation is, of course, the governing equation for a weakly nonlinear, weakly dispersive system with no dissipation. Thus this equation and the perturbed KdV or nonlinear Schrödinger equation, which also describes a weakly nonlinear but strongly dispersive system, are not of primary interest to us but are discussed only because researchers may be interested in comparisons with the tremendous volume of work done in these areas by others, and because the solutions of these as initial value problems were the first important contributions to the theory of solitons.

If the equation is written in the form

$$u_t + \alpha u u_x + \beta u_{xxx} = 0$$

scaling of the variables allows choosing different α, β. We have, as is usual, taken $\alpha = -6$, $\beta = 1$. Many treatments let $\alpha = \beta = 1$. The exact solution has been obtained by exact linearization of the initial-value problem with the initial condition $u(x,0) = f(x)$ with $-\infty < x < \infty$ and $f(x)$ rapidly approaching zero as $|x| \to \infty$.

In many cases of interest for the KdV equation, we may have mixed initial boundary value problems rather than pure initial conditions. For example, we may consider solution in $0 < x < \ell$ for $t \geq 0$. Then, in addition to the initial condition $u(x,0)$, we need boundary conditions at $x = 0$ and $x = \ell$.

References

Adomian, G., *Nonlinear Stochastic Operator Equations* , Academic Press, 1986.

Bonzani, I., "Analysis of Stochastic Van der Pol Oscillators Using the Decomposition Method," *Complex and Distributed Systems: Analysis, Simulation and Control* , IMACS, 1986, 163-168.

Rach, R., "A Convenient Computational Form for the Adomian Polynomials," *J. Math. Anal. and Applic.* 102, 2, 1984, 415-419.

CHAPTER 8

The Benjamin-Ono Equation

The Benjamin-Ono equation can be given in the form

$$u_t = uu_x + Hu_{xx} \tag{8.1}$$

where H denotes the Hilbert transform defined by

$$Hf(x) = \int_{-x}^{\infty} f(\varepsilon)/\varepsilon - x \, d\varepsilon$$

Let $L = \partial/\partial t$ and $L_x = \partial^2/\partial x^2$. Rewrite (8.1) as

$$L_t u = HL_x u + uu_x$$

$$L_t^{-1} L_t u = L_t^{-1} HL_x u + L_t^{-1} uu_x$$

$$u = u(x,0) + L_t^{-1} HL_x u + L_t^{-1} uu_x$$

Solving for the other linear term in (8.1), we have the two equations

$$HL_x u = L_t u - uu_x$$

$$L_x u = H^{-1} L_t u - H^{-1} uu_x$$

$$u = u(0,t) + tu_t(0,t) + L_x^{-1} H^{-1} L_t u - L_x^{-1} H^{-1} uu_x$$

Adding the equations for u

$$\begin{aligned} u = (1/2)\{&[u(x,0) + u(0,t) + tu_t(0,t)] \\ &+ L_t^{-1} HL_x u + L_t^{-1} uu_x \\ &+ L_x^{-1} H^{-1} L_t u - L_x^{-1} H^{-1} uu_x\} \end{aligned}$$

We identify

$$u_0 = (1/2)\{u(x,0) + u(0,t) + tu_t(0,t)\}$$

if $u_t(0,t)$ is known. If other conditions are specified, we have $\alpha + \beta t$ for the last two terms and evaluate them accordingly. Now writing the nonlinear term uu_x in terms of the A_n polynomials, we have

$$u_{n\geq 1} = (1/2)\{L_t^{-1} HL_x u_n + L_t^{-1} A_n + L_x^{-1} H^{-1} L_t u_n - L_x^{-1} H^{-1} A_n\}$$

The sum of the components from 0 to ∞ is the solution. We expect to have a practical solution to sufficient accuracy for an n-term expansion for relatively low n. We evaluate the ϕ_n for the two equations then combine.

CHAPTER 9

The Sine-Gordon Equation

As a physical application of interest in current research, we consider the sine-Gordon equation $\partial^2 u/\partial x \partial t = \sin u$. If we let $L_x = \partial/\partial x$, $L_t = \partial/\partial t$, $N(u) = \sin u$, we can write either

$$L_x L_t u = N(u) \tag{9.1}$$

$$L_t L_x u = N(u) \tag{9.2}$$

Operating on (9.1) first with the integral operator L_x^{-1} then with L_t^{-1}, we obtain

$$u = u(x,0) + u(0,t) + L_t^{-1} L_x^{-1} N(u) \tag{9.3}$$

(We first get $L_t u = L_t u(0,t) + L_x^{-1} N(u)$ then the result in (9.3) assuming $u(0,0) = 0$.) Now operating on (9.2) with L_t^{-1}, then with L_x^{-1}, we obtain

$$u = u(0,t) + u(x,0) + L_x^{-1} L_t^{-1} N(u) \tag{9.4}$$

Adding (9.3) and (9.4) and dividing by two to get a single equation for u, as discussed in (1986),

$$u = u(0,t) + u(x,0) + (1/2)(L_x^{-1} L_t^{-1} + L_t^{-1} L_x^{-1})N(u)$$

We assume a decomposition of the desired solution u into components u_n to be determined, i.e., $u = \sum_{n=0} u_n$, with u_0 identified as

$$u_0 = u(0,t) + u(x,0) \tag{9.5}$$

Nonlinearities, in this case $N(u) = \sin u$, are expressed by $Nu = \sum_{n=0}^{\infty} A_n(u_0, u_1, \ldots u_n)$, where the A_n are generated for the specific nonlinearity. We now have

$$u = u_0 + (1/2)[L_x^{-1} L_t^{-1} + L_t^{-1} L_x^{-1}]\sum_{n=0}^{\infty} A_n \tag{9.6}$$

where u_0 is given by (9.5). Now for any $n > 0$,

$$u_{n+1} = (1/2)[L_x^{-1} L_t^{-1} + L_t^{-1} L_x^{-1}]A_n$$

For $N(u) = f(u) = \sin u$, we have (1986)

$$A_0 = \sin u_0$$

$$A_1 = u_1 \cos u_0$$

$$A_2 = -(1/2)u_1^2 \sin u_0 + u_2 \cos u_0$$

Verifying the above using the explicit forms in (1986):

$$A_1 = c(1,1)h_1$$

$$A_2 = c(1,2)h_1 + c(2,2)h_2$$

$$A_3 = c(1,3)h_1 + c(2,3)h_2 + c(3,3)h_3$$

$$A_4 = c(1,4)h_1 + c(2,4)h_2 + c(3,4)h_3 + c(4,4)h_4$$

$$A_5 = c(1,5)h_1 + c(2,5)h_2 + c(3,5)h_3 + c(4,5)h_4 + c(5,5)h_5$$

We have from $f(u) = \sin u$

$$h_1 = (\cos u_0)$$

$$h_2 = (-\sin u_0)$$

$$h_3 = (-\cos u_0)$$

$$h_4 = (\sin u_0)$$

$$h_5 = (\cos u_0)$$

Hence

$$A_1 = u_1 \cos u_0$$

$$A_2 = u_2 \cos u_0 + (u_1^2/2!)(-\sin u_0)$$

$$A_3 = u_3 \cos u_0 + u_2u_1(-\sin u_0) + (u_1^3/3!)(-\cos u_0)$$

$$A_4 = u_4 \cos u_0 + u_3u_1(-\sin u_0) + (u_2u_1^2/2!)(-\cos u_0) \qquad (9.7)$$

$$+ (u_1^4/4!) \sin u_0$$

$$A_5 = u_5 \cos u_0 - u_4u_1(-\sin u_0) + (u_3u_1^2/2!)(-\cos u_0)$$

$$+ (u_1^3u_2/3!) \sin u_0 + (u_1^5/5!) \cos u_0$$

etc. Thus since u_0 is known, (9.5) and (9.6) provide the solution $u = u_0 + u_1 + \ldots$ where

$$u_0 = [u(x,0) + u(o,t)]$$

$$u_1 = (1/2)[L_t^{-1} L_x^{-1} + L_x^{-1} L_t^{-1}]A_0$$

$$u_2 = (1/2)[L_t^{-1} L_x^{-1} + L_x^{-1} L_t^{-1}]A_1$$

$$\vdots$$

$$u_n = (1/2)[L_t^{-1} L_x^{-1} + L_x^{-1} L_t^{-1}]A_{n-1}$$

with the A_n in (9.7). Thus the u_n are determinable, and we write ϕ_n as an n-term approximation approaching u in the limit as $n \to \infty$.

The convergence and rate of convergence are extremely difficult matters to discuss. We are considering problems which are physical and have solutions. To obtain these solutions, the given data must result in a well-posed problem. Arbitrary assignments of initial data can result in no solution. Such data must be consistent with the solution. In many cases we have an analytic solution in which dependences are easily seen and sometimes even a closed-form solution is obtained. Consider also the covariant form

$$u_{xx} - u_{tt} = \sin u$$

or

$$L_x u - L_t u = \sin u$$

where $L_x = \partial^2/\partial x^2$ and $L_t = \partial^2/\partial t^2$, then we have

$$L_x u = \sin u + L_t u$$

$$L_t u = - \sin u + L_x u$$

or

$$u = \alpha + \beta x + L_x^{-1} \sin u + L_x^{-1} L_t u$$

$$u = \Upsilon + \delta t - L_t^{-1} \sin u + L_t^{-1} L_x u$$

Finally

$$u = (1/2)\{\alpha + \beta x + \Upsilon + \delta t\}$$
$$+ (1/2)\{L_x^{-1} - L_t^{-1}\}\sin u + (1/2)\{L_x^{-1} L_t + L_t^{-1} L_x\}u$$

Thus

$$u_0 = (1/2)\{\alpha + \beta x + \Upsilon + \delta t\}$$

$$u_{n\geq 1} = (1/2)\{L_x^{-1} - L_t^{-1}\}\sin u + (1/2)\{L_x^{-1} L_t + L_t^{-1} L_x\}u_n$$

where the constants in the u_0 term satisfy given conditions. This form requires two additional conditions, $\alpha = u(0,t)$, $\Upsilon = u(x,0)$. If additional conditions $u_x(0,t)$ and $u_t(x,0)$ are given, $\beta = u_x(0,t)$, $\delta = u_t(x,0)$. If other conditions are given on a finite or infinite interval, β and δ will be evaluated accordingly. If $u_t(x,0) = u_x(0,t) = 0$, and we specify $u(x,0) = f(x)$ and $u(0,t) = g(t)$, the solution is determined.

Exercise:
Let $f(x) = \sec^2 x$ and $g(t) = \sec^2(-t)$ and find u.

Exercise:

Consider the transformed form of the sine-Gordon equation $\phi_{xx} - \phi_{tt} = \sin \phi$ applicable to a basic model of the Josephson junction when the nonlinear dissipation term and the external bias current β are neglected. If these are included as well, we have

$$\phi_{xx} - \phi_{tt} = \alpha|\phi_t|\phi_t = \sin \phi - \beta$$

If α, β are given constants and we assume the conditions $\phi_x(0,t) = \phi_x(\ell,t) = 0$ and $\phi(x,0)$ and $\phi(0,t)$ are given, find the solution.

Questions of existence and uniqueness or a general proof of convergence cannot be completely asnwered at this time and remain to be studied. Solutions, once obtained, are verifiable by substitution at any level of approximation. To see this, consider a simple example: $y^{\cdot} + y^2 = x(t)$ where $x(t) = t^2 + 1$ and $y(0) = 0$. For the two-term approximation ϕ_2 obtained by decomposition, we get $\phi_2 = (t^3/3 + t) - (t^7/63 + 2t^2/15 + t^3/3)$. If we use $y \simeq \phi_2$, we have:

$$y \simeq t^2 + 1 - t^6/9 - 2t^4/3 - t^2$$

Thus $y + y^2 = t^2 + 1$ verifying the approximation. Note that in substituting for y^2, we use only y_0 here, not $y_0 + y_1$. Approximating the derivative to y_1 requires only A_0 since y depends only on $A_0(y_0)$. Thus $\sum_{i=0}^{\infty} y_i^{\cdot} + \sum_{i=0}^{n-1} A_i = x$. As $n \to \infty$, we have the original equation.

We are concerned with solving physical systems and need not consider unbounded or pathological inputs. There is now no doubt that the decomposition method provides a very useful solution technique for a wide class of systems. While it is possible to construct an example which produces no solution, this proves nothing whatsoever. The results depend on the particular initial/boundary conditions assumed or given in any physical problem. Depending on the given data, solutions may exist and be unique or not. If the conditions are physically correct, we get a solution which is in a series form with some region of convergence. Numerical evaluation of the solution and quantitative comparisons can then be made. We are well aware that solutions to some frontier problems here may be viewed as purely formal, and we welcome those with computer access (which we do not have) to go further. Our effort here is necessarily confined for the present to show a procedure which, properly used, should yield a physically more correct solution than conventional methods requiring linearization, perturbation, etc. as well as avoiding massive computations in many cases.

A simple computation with $Ly + Ny = x(t)$ can be instructive. For simplicity, let $L = d/dt$ and $Ny = y^2$ with a zero initial condition. We get $y_1 = -L^{-1}y^2$ where L^{-1} is the definite integral from 0 to t. Now

$$|y_1| = M^3t^3/3!$$

if M is the bound for $x(t)$ in $[0,T]$ and

$$|y_2| = M^5t^5/5!$$

etc.

so that $|y_{n+1}| \to 0$ for t in $[0,T]$ where T is finite. For ϕ_n we have $\phi_n = y_0 - L^{-1} \sum_{i=0}^{n-1} A_i$ and $\phi_{n+1} = y_0 - L^{-1} \sum_{n=0}^{n} A_i$. Hence,

$$|\phi_{n+1} - \phi_n| = -|L^{-1} A_n| = -|y_{n+1}| \to 0$$

Numerical results have shown a high degree of accuracy for rather low n in many problems and sometimes the series can be summed. In (1983) we developed error estimates to show the effect of calculating more terms. However, when numerical results are obtained, one sees the approach to a stable solution for the desired number of decimal places.

The effect of linearization is easily determined in a specific problem by calculating with $Ny = \sum_{n=0}^{\infty} A_n$ and comparing with the result for a linearized Ny.

Of course, a rigorous mathematical framework should be sought but may take a long time, and there is a fertile field here for research as well as the possibility of much needed insights into application.

References

Adomian, G., *Nonlinear Stochastic Operator Equations*, Academic Press, 1986.

Adomian, G., *Stochastic Systems*, Academic Press, 1983.

CHAPTER 10

The Nonlinear Schrödinger Equation and the Generalized Schrödinger Equation

10.1 Nonlinear Schrödinger Equation

Here, we consider the equation

$$iu_t + 2u|u|^2 + u_{xx} = 0 \tag{10.1.1}$$

which we will write

$$iL_t u + Nu + L_x u = 0$$

where $Nu = 2u|u|^2$, $L_t = \partial/\partial t$, $L_x = \partial^2/\partial x^2$. Solving for $L_x u$ we have

$$\begin{aligned} L_x u &= -iL_t u - Nu \\ u &= \alpha + \beta x - iL_x^{-1} L_t u - L_x^{-1} Nu \end{aligned} \tag{10.1.2}$$

Solving for $L_t u$ we have

$$\begin{aligned} L_t u &= iNu + iL_x u \\ u &= \Upsilon + iL_t^{-1} Nu + iL_t^{-1} L_x u \end{aligned} \tag{10.1.3}$$

Add (10.1.2) and (10.1.3), then

$$\begin{aligned} u &= (1/2)\{\alpha + \beta x + \Upsilon\} \\ &+ (1/2)\{iL_t^{-1} Nu - L_x^{-1} Nu\} \\ &+ (1/2)\{iL_t^{-1} L_x - L_x^{-1} L_t\}u \end{aligned} \tag{10.1.4}$$

Thus

$$u_0 = (1/2)\{\alpha + \beta x + \Upsilon\}$$

$$u_{n\geq 1} = (1/2)\{iL_t^{-1} - L_x^{-1}\}\sum_{n=0}^{\infty} A_n \qquad (10.1.5)$$

$$+ (1/2)\{iL_t^{-1} L_x - iL_x^{-1} L_t\}u_n$$

The constants α, β, Υ are evaluated from the given conditions

$$\Upsilon = u(x,0)$$

$$\alpha = u(0,t)$$

$$\beta = u_x(0,t)$$

If we have mixed initial/boundary conditions, β changes accordingly. If we assume that $u \to 0$ as $x \to \infty$, $\beta = 0$.

10.2 Generalized Schrödinger Equation

Consider $\nabla^2 u + u_t + Nu = g$. We can write this immediately in the form:

$$[L_x + L_y + L_z]u + L_t u + Nu = g$$

where $L_x = \partial/\partial x$, $L_y = \partial/\partial y$, $L_z = \partial/\partial z$, $L_t = \partial/\partial t$. We obtain four equations by solving alternately for the four linear terms. Thus

$$L_x u = g - L_y u - L_z u - L_t u - Nu$$

$$L_y u = g - L_z u - L_t u - L_x u - Nu$$

$$L_z u = g - L_t u - L_x u - L_y u - Nu$$

$$L_t u = g - L_x u - L_y u - L_z u - Nu$$

Defining the usual inverses to act on both sides

$$u = \Upsilon_1 + \Upsilon_2 x + L_x^{-1} g - L_x^{-1} L_y u - L_x^{-1} L_z u - L_x^{-1} L_t u - L_x^{-1} Nu$$

$$u = \Upsilon_3 + \Upsilon_4 y + L_y^{-1} g - L_y^{-1} L_z u - L_y^{-1} L_t u - L_y^{-1} L_x u - L_y^{-1} Nu$$

$$u = \Upsilon_5 + \Upsilon_6 z + L_z^{-1} g - L_z^{-1} L_t u - L_z^{-1} L_x u - L_z^{-1} L_y u - L_z^{-1} Nu$$

$$u = \Upsilon_7 + L_t^{-1} g - L_t^{-1} L_x u - L_t^{-1} L_y u - L_t^{-1} L_z u - L_t^{-1} Nu$$

Let $u_0 = (1/4)\{\Upsilon_1 + \Upsilon_2 x + \Upsilon_3 + \Upsilon_4 y + \Upsilon_5 + \Upsilon_6 z + \Upsilon_7$

$$+ [L_x^{-1} + L_y^{-1} + L_z^{-1} + L_t^{-1}]g\}$$

Then

$$u = u_0 - (1/4)\{[L_x^{-1} L_y + L_x^{-1} L_z + L_x^{-1} L_t]$$

$$+ [L_y^{-1} L_z + L_y^{-1} L_t + L_y^{-1} L_x] + [L_z^{-1} L_t + L_z^{-1} L_x + L_z^{-1} L_y]$$

$$+ [L_t^{-1} L_x + L_t^{-1} L_y + L_t^{-1} L_z]\}u - (1/4)\{L_x^{-1} + L_y^{-1} + L_z^{-1} + L_t^{-1}\}Nu$$

Now replacing u by $\sum_{n=0}^{\infty} u_n$ and Nu by $\sum_{n=0}^{\infty} A_n$, we can calculate u_n for $n = 1,2,3,\ldots$ in terms of u_0 by

$$u_n = -(1/4)\{L_x^{-1}[L_y + L_z + L_t] + L_y^{-1}[L_z + L_t + L_x]$$

$$+ L_z^{-1}[L_t + L_x + L_y] + L_t^{-1}[L_x + L_y + L_z]\}u_{n-1}$$

$$- (1/4)\{L_x^{-1} + L_y^{-1} + L_z^{-1} + L_t^{-1}\}A_n$$

for $n = 1,2,3,\ldots$. Thus it remains only to evaluate the constants in u_0 from the given initial/boundary conditions.

10.3 Schrödinger's Equation with a Quartic Potential

Consider the equation

$$(-\hbar^2/2m)d^2\psi/dx^2 + (1/2)ax^4\psi = E_1\psi$$

for a point particle in a quartic potential $V(x) = (1/2)\alpha x^4$. Let $\alpha = -ma/\hbar^2$ and $\beta = 2mE_1/\hbar^2$. Then the equation becomes

$$d^2\psi/dx^2 + \alpha x^4\psi + \beta\psi = 0$$

Assume that the energy of each eigenstate has been determined as discussed in references (1986). Now, if $L = d^2/dx^2$, we can write

$$L\psi + \alpha x^4\psi + \beta\psi = 0$$

or

$$L\psi = -\alpha x^4\psi - \beta\psi$$

If we denote by Φ the solution of $L\psi = 0$, we have

$$\psi = \Phi - L^{-1}(\alpha x^4 + \beta)\psi$$

so that, using the decomposition method

$$\psi_0 = \Phi$$

$$\psi_{n+1} = -L^{-1}(\alpha x^4 + \beta)\psi_n$$

for $n \geq 0$. Then $\psi = \sum_{n=0}^{\infty} \psi_n$ is the solution.

In three dimensions with $V(\bar{x}) = 1/2\, \alpha|\bar{x}|^4\psi$, writing $L_x = \partial^2/\partial x^2$, $L_y = \partial^2/\partial y^2$, $L_z = \partial^2/\partial z^2$, letting Φ_x, Φ_y, Φ_z denote the solutions of $L_x\psi = 0$, $L_y\psi = 0$, $L_z\psi = 0$ respectively, the decomposition procedure yields:

$$[L_x + L_y + L_z]\psi + \alpha x^4\psi + \beta\psi = 0$$

Solving for the linear operator terms

$$L_x\psi = -L_y\psi - L_z\psi - \alpha x^4\psi - \beta\psi$$

$$L_y\psi = -L_x\psi - L_z\psi - \alpha x^4\psi - \beta\psi$$

$$L_z\psi = -L_x\psi - L_y\psi - \alpha x^4\psi - \beta\psi$$

With the inversions

$$L_x^{-1} L_x\psi = -(L_x^{-1} L_y + L_x^{-1} L_z)\psi - L_x^{-1} \alpha x^4\psi - L_x^{-1} \beta\psi$$

$$L_y^{-1} L_y\psi = -(L_y^{-1} L_x + L_y^{-1} L_z)\psi - L_y^{-1} \alpha x^4\psi - L_y^{-1} \beta\psi$$

$$L_z^{-1} L_z\psi = -(L_z^{-1} L_x + L_z^{-1} L_y)\psi - L_y^{-1} \alpha x^4\psi - L_z^{-1} \beta\psi$$

and so, we have the three equations

$$\psi = \Phi_x - (L_x^{-1} L_y + L_x^{-1} L_z)\psi - L_x^{-1} \alpha x^4\psi - L_x^{-1} \beta\psi$$

$$\psi = \Phi_y - (L_y^{-1} L_x + L_y^{-1} L_z)\psi - L_y^{-1} \alpha x^4\psi - L_y^{-1} \beta\psi$$

$$\psi = \Phi_z - (L_z^{-1} L_x + L_z^{-1} L_y)\psi - L_z^{-1} \alpha x^4\psi - L_z^{-1} \beta\psi$$

The Φ_x, Φ_y, Φ_z are completely specified by initial/boundary conditions. Now adding and dividing by three, we get a single equation where $\psi_0 = (1/3)(\Phi_x + \Phi_y + \Phi_z)$

$$\psi = \psi_0 - (1/3)[L_x^{-1} L_y + L_x^{-1} L_z + L_y^{-1} L_x + L_y^{-1} L_z + L_z^{-1} L_x + L_z^{-1} L_y]\psi$$
$$- (1/3)[L_x^{-1} + L_y^{-1} + L_z^{-1}]\alpha x^4\psi$$
$$- (1/3)[L_x^{-1} + L_y^{-1} + L_z^{-1}]\beta\psi$$

If, for convenience in writing, we let

$$K = -(1/3)[L_x^{-1} L_y + L_x^{-1} L_z + L_y^{-1} L_x + L_y^{-1} L_z + L_z^{-1} L_x + L_z^{-1} L_y]$$

$$G = -(1/3)[L_x^{-1} + L_y^{-1} + L_z^{-1}]$$

we have

$$\psi = \psi_0 + K\psi + G\alpha x^4\psi + G\beta\psi$$

Letting $\psi = \sum_{n=0}^{\infty} \psi_n$, we have ψ_0 given and

$$\psi_{n+1} = K\psi_n + G(\alpha x^4 + \beta)\psi_n$$

for $n \geq 0$, yielding the components of ψ.

References

Adomian, G., *Nonlinear Stochastic Operator Equations* , Academic Press, 1986.

Adomian, G., "A General Approach to Solution of Partial Differential Equation Systems," *Comput. Math. Applic.* , 13, 9-11, 1987, 741-747.

Adomian, G., "A New Approach to the Efinger Model for a Nonlinear Quantum Theory for Gravitating Particles," *Found of Phys.* , 17, 4,1987, 419-423.

Efinger, H., "On the Theory of Certain Nonlinear Schrödinger Equations with Nonlocal Interaction," *Nuovo Cimento* , B 80, 260, 1984.

Efinger, H. and Grösse, H. "On Bound State Solutions for Certain Nonlinear Schrödinger Equations," *Lett. Math. Phys* ., 8, 91, 1984.

CHAPTER 11

Nonlinear Plasmas

Until recently most or all of work in the plasma area has utilized linear models and linear analysis for reasons of mathematical tractability. In some cases, however, "perturbations" are either large or start off small and then grow rapidly as in transition to turbulence (which is a strongly nonlinear and stochastic phenomena), and possibly in the formation of galaxies and systems of galaxies.

Calculation of the response of an interacting electron stream to electrostatic perturbation is usually done using the self-consistent field technique (1973; 1959; 1959). Use of the self-consistent field method requires taking the expectation value of the equation of motion, whereas one should solve the equation first as in the decomposition method and then average. The usual method is a typical hierarchy procedure which necessarily neglects interaction terms, simply replacing an average of a product by a product of averages. In weakly coupled plasmas, this may be a sufficiently accurate approximation; however, the error in doing so for the general case has been well established (1983; 1979). It is, in fact, a perturbation method in which the instantaneous field experienced by an electron does not deviate much from the average value. It has been shown that a better, and certainly more physically realistic method of solution is possible now.

The total charge density consists of two parts - an external charge density ρ_{ext} and an induced charge density ρ_{ind} due to the nonuniformity of the plasma. The electrostatic potential in a single-component plasma can be calculated from Poisson's equation

$$\nabla^2\phi = -\rho_{ext} + \alpha(e^{-\beta\phi} - 1) \qquad (11.1)$$

where the second term on the right is the induced charge ρ_{ind}. Thus, we have a nonlinear differential equation for ϕ. At this point it is customary to make various approximations, e.g., ρ_{ext} is assumed to be small so that the induced potential will also be small ($\beta\phi << 1$). One can then write a linearized equation. Instead of doing this we will write (11.1) in the standard form of (1983) as

$$\nabla^2\phi - \alpha e^{-\beta\phi} = -\rho_{ext} - \alpha \tag{11.2}$$

for $F\phi \equiv L\phi + N\phi = g$ in our standard notation, where $L = \nabla^2$, $N\phi = -\alpha e^{-\beta\phi}$, and $g = -\rho_{ext} - \alpha$. Let $L = L_x + L_y + L_z$ where $L_x = \partial^2/\partial x^2$, $L_y = \partial^2/\partial y^2$, $L_z = \partial^2/\partial z^2$ (1985). Then we have

$$[L_x + L_y + L_z]\phi + N\phi = g$$

We solve for $L_x\phi$, $L_y\phi$, $L_z\phi$ in turn as discussed in (1985) so that

$$L_x\phi = g - L_y\phi - L_z\phi - N\phi$$

$$L_y\phi = g - L_z\phi - L_x\phi - N\phi$$

$$L_z\phi = g - L_x\phi - L_y\phi - N\phi$$

The inverses L_x^{-1}, L_y^{-1}, L_z^{-1} are defined (1983; 1986) as two-fold integrations. If we indicate the homogeneous solutions by Φ_x, Φ_y, Φ_z (i.e., $L_x\Phi_x = 0$, etc.)

$$\phi = \Phi_x + L_x^{-1} g - L_x^{-1} L_y\phi - L_x^{-1} L_z\phi - L_x^{-1} N\phi$$

$$\phi = \Phi_y + L_y^{-1} g - L_y^{-1} L_z\phi - L_y^{-1} L_x\phi - L_y^{-1} N\phi$$

$$\phi = \Phi_z + L_z^{-1} g - L_z^{-1} L_x\phi - L_z^{-1} L_y\phi - L_z^{-1} N\phi$$

Adding and dividing by three,

$$\begin{aligned} \phi = {} & (1/3)\{(\Phi_x + \Phi_y + \Phi_z) + (L_x^{-1} + L_y^{-1} + L_z^{-1})g\} \\ & - (1/3)\{L_x^{-1} L_y + L_y^{-1} L_z + L_z^{-1} L_x + L_x^{-1} L_z + L_y^{-1} L_x + L_z^{-1} L_y\}\phi \\ & - (1/3)\{L_x^{-1} + L_y^{-1} + L_z^{-1}\}N\phi \end{aligned}$$

We define the entire first line above as ϕ_0. Define the entire second line as $-(L^{-1}R)\phi$ and the third line as $L^{-1}N\phi$; then we have the convenient form

$$\phi = \phi_0 - (L^{-1}R)\phi - L^{-1}(N\phi) \qquad (11.3)$$

For ϕ we substitute $\phi = \sum_{n=0}^{\infty} \phi_n$ and $N\phi$ is developed in our A_n polynomials. Thus

$$\sum_{n=0}^{\infty} \phi_n = \phi_0 - (L^{-1}R) \sum_{n=0}^{\infty} \phi_n - L^{-1} \sum_{n=0}^{\infty} A_n \qquad (11.4)$$

The A_n, of course, are developed for the particular nonlinearity present (1983; 1986). The components of ϕ are:

$$\begin{aligned}
\phi_0 &= (1/3)\{(\Phi_x + \Phi_y + \Phi_z) + (L_x^{-1} + L_y^{-1} + L_z^{-1})g\} \\
\phi_1 &= - L^{-1}R\phi_0 - L^{-1}A_0 \\
\phi_2 &= - L^{-1}R\phi_1 - L^{-1}A_1 \\
\phi_3 &= - L^{-1}R\phi_2 - L^{-1}A_2 \\
&\vdots
\end{aligned} \qquad (11.5)$$

The A_n have been previously defined (1984; 1986) for $N\phi = \alpha e^{-\beta\phi}$. They are:

$$\begin{aligned}
A_0 &= \alpha e^{-\beta\phi_0} \\
A_1 &= \alpha e^{-\beta\phi_0}[-\beta\phi_1] \\
A_2 &= \alpha e^{-\beta\phi_0}[-\beta\phi_2 + (\beta^2/2!)\phi_1^2] \\
A_3 &= \alpha e^{-\beta\phi_0}[-\beta\phi_3 + \beta^2\phi_1\phi_2 - (\beta^3/3!)\phi_1^3] \\
A_4 &= \alpha e^{-\beta\phi_0}[-\beta\phi_4 + \beta^2\{(\phi_2/2!) + \phi_1\phi_3\} \\
&\quad - (\beta^3\phi_1^2\phi_2/2!) + (\beta^4\phi_1^4/4!)] \qquad (11.6) \\
A_5 &= \alpha e^{-\beta\phi_0}[-\beta\phi_5 + \beta^2\{\phi_2\phi_3 + \phi_1\phi_4\} \\
&\quad - \beta^3\{(\phi_1\phi_2^2/2!) + (\phi_1^2\phi_3/2!)\} \\
&\quad + (\beta^4\phi_1^3\phi_2/3!) + (\beta^5\phi_1^5/5!)]
\end{aligned}$$

$$A_6 = \alpha e^{-\beta\phi_0}[-\beta\phi_6 + \beta^2\{(\phi_3^2/2!) + \phi_2\phi_4 + \phi_1\phi_5\}$$

$$- \beta^3\{(\phi_2^3/3!) + \phi_1\phi_2\phi_3 + \phi_1^2\phi_4/2!\}$$

$$+ \beta^4\{(\phi_1^2/2!)(\phi_2^2/2!) + (\phi_1^3/3!)\phi_3\} - \beta^5(\phi_1^4/4!)\phi_2 + \beta_1(\phi_1^6/6!)]$$

$$A_7 = \alpha e^{-\beta\phi_0}[-\beta\phi_7 + \beta^2\{\phi_3\phi_4 + \phi_2\phi_5 + \phi_1\phi_6\}$$

$$- \beta^3\{(\phi_2^2/2!)\phi_3 + \phi_1(\phi_3^2/2!) + \phi_1\phi_2\phi_4$$

$$+ (\phi_1^2/2!)\phi_5\} + \beta^4\{\phi_1(\phi_2^3/3!) + (\phi_1^2/2!)\phi_2\phi_3$$

$$+ (\phi_1^3/3!)\phi_4\} - \beta^5\{(\phi_1^3/3!)(\phi_2^2/2!) + (\phi_1^4/4!)\phi_3\}$$

$$+ \beta^6(\phi_1^5/5!)\phi_2 - \beta^7(\phi_1^7/7!)]$$

$$A_8 = \alpha e^{-\beta\phi_0}[-\beta\phi_8 + \beta^2\{(\phi_4^2/2!) + \phi_3\phi_5 + \phi_2\phi_6 + \phi_1\phi_7\}$$

$$- \beta^3\{\phi_2(\phi_3^2/2!) + (\phi_2^2/2!)\phi_4 + \phi_1\phi_3\phi_4$$

$$+ \phi_1\phi_2\phi_5 + (\phi_1^2/2!)\phi_6\}$$

$$+ \beta^4\{(\phi_2^4/4!) + \phi_1(\phi_2^2/2!)\phi_3 + (\phi_1^2/2!)(\phi_3^2/2!)$$

$$+(\phi_1^2/2!)\phi_2\phi_4 + (\phi_1^3/3!)\phi_5\}$$

$$- \beta^5\{(\phi_1/2!)(\phi_2^3/3!) + (\phi_1^3/3!)\phi_2\phi_3 + (\phi_1^4/4!)\phi_4\}$$

$$+ \beta^6\{(\phi_1^4/4!)(\phi_2^2/2!) + (\phi_1^5/5!)\phi_3\}$$

$$- \beta^7(\phi_1^6/6!)\phi_2 + \beta^8(\phi_1^8/8!)]$$

$$A_9 = \alpha e^{-\beta\phi_0}[-\beta\phi_9 + \beta^2\{\phi_4\phi_5 + \phi_3\phi_6 + \phi_2\phi_9 + \phi_1\phi_8\}$$

$$- \beta^3\{(\phi_3^3/3!) + \phi_2\phi_3\phi_4 + (\phi_2^2/2!)\phi_5 + \phi_1(\phi_4^2/2!)$$

$$+ \phi_1\phi_3\phi_5 + \phi_1\phi_2\phi_6 + (\phi_1^2/2!)\phi_7\}$$

$$+ \beta^4\{(\phi_2^3/3!)\phi_3 + \phi_1\phi_2(\phi_3^2/2!) + \phi_1(\phi_2^2/2!)\phi_4$$

$$+(\phi_1^2/2!)\phi_3\phi_4 + (\phi_1^2/2!)\phi_2\phi_5 + (\phi_1^3/3!)\phi_6\}$$

$$- \beta^5(\phi_1^4(\phi_2/4!) + (\phi_1^2/2!)(\phi_2^2/2!)\phi_3$$

$$+ (\phi_1^3/3!)(\phi_3^2/2!) + (\phi_1^3/3!)\phi_2\phi_4 + (\phi_1^4/4!)\phi_5\}$$

$$+ \beta^6\{(\phi_1^3/3!)(\phi_2^3/3!) + (\phi_1^4/4!)\phi_2\phi_3 + (\phi_1^5/5!)\phi_4\}$$

$$- \beta^7\{(\phi_1^5/5!)(\phi_2^2/2!) + (\phi_1^6/6!)\phi_3\}$$

$$+ \beta^8(\phi_1^7/7!)\phi_2 - \beta^9(\phi_1^9/9!)]$$

which is sufficient to give us the ten-term approximation ϕ_{10}. (More are easily calculated if the problem warrants it, but we expect a very rapidly damped oscillating convergence from results discussed previously - particularly in (1984;1986.)

Since $\beta\phi$ is generally assumed to be small ($\beta\phi<<1$), $e^{-\beta\phi} - 1$ becomes $-\beta\phi$ resulting in a linear equation. The decomposition method solves the equation (11.1) without linearizing assumptions, perturbative methods, or truncations and the complete solution is given by ϕ_n.

A complete and accurate solution can now be obtained for any specific initial/boundary conditions for (11.1) and values of α, β, ρ_{ext} in a rapidly convergent series (1984) to any desired approximation.

References

Adomian, G. "Convergent Series Solution of Nonlinear Equations," *J. Comput. and Appl. Math,* 11 , 2, 1984, 225-230.

Adomian, G. *Nonlinear Stochastic Operator Equations* , Academic Press, 1986.

Adomian, G., *Stochastic Systems* , Academic Press, 1983.

Adomian, G. and R. Rach, "Nonlinear Plasma Response," *J. Math.Anal. and Appl.,* 111, 1, 1985, 114-118.

Bellomo, N. and R. Monaco, "A Comparison between Adomian's Decomposition Methods and Perturbation Techniques for Nonlinear Random Differential Equations," *J. Math. Anal. and Applic.* , 110, 2, 1985, 495-502.

Bellomo, N., G. Rizzi, and E. Cafaro, "On the Mathematical Modelling of Physical Sytems by Stochastic Differential Equations," *Proc. X IMACS Conf.*, Montreal, 1982, 322-325.

Ehrenreich, H. and M. H. Cohen, *Phys. Rev.* 115, 1959, 786.

Goldstone, J. and K. Gottfried, *Nuovo Cimento*, 10, 13, 1959, 849.

Platzman, P. M. and P. A. Wolff, *Waves and Interactions in Solid State Plasmas*, Academic Press, 1973.

Rach, R.,"A Convenient Computational Form for the Adomian Polynomials," *J. Math. Anal. and Appl.*, 102, 2, 1984, 415-419.

CHAPTER 12

The Tricomi Problem

The Tricomi equation (1968; 1923) is a fundamental mathematical model for the solution of the problem of defining the flow-field around wing shapes at transonic speeds. This equation can be classified as a "mixed type" equation, which is elliptic in the positive half-space ($y > 0$) of a cartesian plane with axes x, y and hyperbolic in the negative half-space ($y < 0$), the boundary $y = 0$ being the parabolic separation line. Analytical solutions of the Tricomi problem are useful in those applications such that the boundary conditions are assigned for transonic wing profiles (1964;1973). In this chapter we propose an analytical solution obtained in (1985) using the decomposition method for partial differential equations (1986).

The Tricomi equation is given in the form:

$$y\phi_{xx} + \phi_{yy} = 0 \tag{12.1}$$

In the analysis of the above problem we shall denote by $\tau(x)$ and $\upsilon(x)$ two functions such that

$$\tau(x) = \phi(x, y{=}0), \qquad \upsilon(x) = \phi_y(x, y{=}0) \tag{12.2}$$

and by $u(y)$ and $v(y)$ the functions such that:

$$u(y) = \phi(x{=}0, y), \qquad v(y) = \phi_x(x{=}0, y) \tag{12.3}$$

Analysis: The analysis will be carried out here only for the problem of determining the function $\phi(x,y)$ in the half-space $x \geq 0$, given the values of the functions u and v. A class of quasi-analytical solutions will be determined. A more detailed analysis appears in (1985) for the complete Tricomi problem. Since the solutions are carried out in the same manner, we restrict attention to this half-space problem.

$$L_x = \partial^2/\partial x^2, \quad L_y = \partial^2/\partial y^2 \tag{12.4}$$

whose inverses exist for every $q = q(x,y) \varepsilon B$ where B is the space of all functions $\phi = \phi(x,y)$ such that:

1) ϕ is a C^2 function for every $(x,y) \varepsilon D$.
2) ϕ is continuous on the boundary of D.
3) ϕ_x and ϕ_y are continuous on the boundary of D.

$$L_x^{-1} q = \int_0^x dr \int_0^r q(s,y)ds, \quad L_y^{-1} q = \int_0^y dr \int_0^r q(x,s)ds \tag{12.5}$$

Equation (12.1) can now be rewritten in one of the two forms:

$$L_y\phi = - yL_x\phi \qquad L_x\phi = - y^{-1} L_y\phi \tag{12.6}$$

Each of them can be formally integrated applying the inverse operators L_y^{-1} to the first one and L_x^{-1} to the second one; then the addition, (and division by two,) provides the following operator equivalent formulation of equation (12.1):

$$\phi(x,y) = \phi_0(x,y) - K(x,y)\phi(x,y) \tag{12.7}$$

where

$$\phi_0(x,y) = (1/2)\{\tau(x) + y\upsilon(x) + u(y) + xv(y)\}$$

and the operator K is defined as follows:

$$K = (1/2)\{L_y^{-1} yL_x + L_x^{-1} (1/y)L_y\}$$

a mixed-type integral-differential operator. The decomposition method can now be applied to writing the solution of (12.7) in the form:

$$\phi(x,y) = \phi_0(x,y) - K \sum_{n=0}^{\infty} \phi_n(x,y) \tag{12.8}$$

from which we identify

$$\phi_{n+1} = - K\phi_n \tag{12.9}$$

and we see that all components are determinable in a sequence of easily computable quadratures, each defined by the preceding one.

Remark: Note that the solution defines a fixed-point formulation of the problem so that the classical Cacciopoli-Banach fixed point theorem can be applied in order to prove the existence and uniqueness of the solutions.

References

Adomian, G., *Nonlinear Stochastic Operator Equations* , Academic Press, 1986.

Adomian, G. and N. Bellomo, "On the Tricomi Problem," *Advances in Hyperbolic Equations* , 3, 1985; also in *Computers and Mathematics with Applications* (Special issue on medicine), 1985.

Ferrari, C. and F. Tricomi, *Transonic Aerodynamics* , Academic Press, 1968.

Morawetz, C., "Non Existence of Transonic Flow Past a Profile," *Comm. Pure Appl. Math.* , 14,1964, 357-367.

Nocilla, S., "Sulla Determinazione del Flusso Transonico Continuo Attorno a Profili Alari Aimmetrici con Curvatura Regolare Senza Incidenza, l'Aerotecnica," *J. Italian Ass. Aeron. & Astron..*, 4 1973, 245-260.

Tricomi, F., "Sulle Equazioni Lineari alle Derivate Parziali di secondo Ordine di Tipo misto," *Memorie Accademia Lincei* , 5, 14, 1923, 133- 247.

Chapter 13

The Initial-Value Problem for the Wave Equation

Consider the hyperbolic partial differential equation

$$\partial^2 u/\partial x^2 = \partial^2 u/\partial t^2 \tag{13.1}$$

for $t > 0$ and $0 < x < \ell$ with $\phi(x,0)$ and $\phi_t(x,0)$ specified for $0 < x < \ell$, and $\phi(0,t)$ and $\phi(\ell, t)$ specified for $t > 0$. Let $L_x = \partial^2/\partial x^2$, $L_t = \partial^2/\partial t^2$. (13.1) is then written

$$L_x\phi = L_t\phi$$

Operating with L_x^{-1} defined as a two-fold integral operator from 0 to x

$$\phi = \phi_x + L_x^{-1} L_t\phi$$

where $\phi = A + Bx$. Similarly operating with L_t^{-1}

$$\phi = \phi_t + L_t^{-1} L_x\phi$$

Thus

$$\phi = (1/2)\{\phi_x + \phi_t\} + (1/2)\,\{L_x^{-1} L_t + L_t^{-1} L_x\}\phi$$

We obtain $\phi_x + \phi_t$ from the initial/boundary conditions; then we define

$$\phi_0 = (1/2)\{\phi_x + \phi_t\}$$

and identify

$$\phi_{n+1} = (1/2)\{L_x^{-1} L_t + L_t^{-1} L_x\}\phi_n \qquad n \geq 0$$

We need only to identify ϕ_0 in terms of the specified conditions for a complete solution; then all components are determined, e.g.

$$\phi_t = \phi(x,0) + t\phi_t(x,0)$$

$$\phi_x = \phi(0,t) + x[\phi(\ell, t) - \phi(0,t)]/\ell$$

The wave equation is usually written

$$\partial^2\psi/\partial t^2 = \alpha^2\, \partial^2\psi/\partial x^2$$

assuming no dissipation or dispersion, small amplitudes, and that nonlinear terms in ψ can be disregarded.

With these assumptions, the form of the equations does not depend on properties of the medium. If the assumptions are not valid, then each medium requires its own characteristic equation.

CHAPTER 14

Nonlinear Dispersive or Dissipative Waves

14.1 Wave Propagation in Nonlinear Media

The behavior of nonlinear dispersive or dissipative waves will now be investigated by decomposition. Thus, consider the solution $\phi(x,t)$ of the nonlinear partial differential equation

$$\partial\phi/\partial t + a\phi^m\partial\phi/\partial x = b\partial^n\phi/\partial x^n \quad (t > 0,\ x > 0) \qquad (14.1.1)$$

where $m = 1,2$ and $n = 2,3,4...$ subject to the initial condition

$$\phi(x,0) = f(x) \qquad (x \geq 0)$$

and boundary conditions

$$\phi(0,t) = g(t)$$

$$\lim_{x\to\infty} \phi(x,t) = 0 \qquad (t \geq 0)$$

where a, b are real constants and $f(x)$ and $g(t)$ are given sufficiently smooth functions of x and t respectively satisfying the condition $f(0) = g(0)$. Oguztoreli, Shuhubi, and Leung have discussed a numerical solution of this problem and have pointed out that if $m = 1$, equation (14.1.1) is the Burger's equation when $n = 2$ and the KdV equation when $n = 3$ (and that it may be called a generalized Burger's equation for even $n \geq 4$ and $m > 1$ or a generalized KdV equation for odd n and the same conditions).

Our objective is an analytic solution which is obtained in a rapidly convergent series form without linearization. Questions of existence and uniqueness are left to discussion elsewhere (1986). Write equation (14.1.1) in the form

$$L_t\phi + aN\phi = bL_x\phi \qquad (14.1.2)$$

where $L_t = \partial/\partial t$, $L_x = \partial^n/\partial x^n$, $N\phi = \phi^m \phi_x$, and $m = 1, 2$ and $n = 2,3,4,\dots$. Solving for the linear terms as discussed in (1986) we get two equations

$$L_x\phi = b^{-1} L_t\phi + ab^{-1}N\phi \tag{14.1.3}$$

$$L_t\phi = bL_x\phi - aN\phi \tag{14.1.4}$$

The inverse operator L_t^{-1} can conveniently be taken as the definite integral from 0 to t. The inverse L_x^{-1} is taken as an indefinite (see 1986) n-fold integration. Thus $L_x^{-1}L_x\phi = \phi + x\alpha(t) + \beta(t)$ if L_x is second-order, and is changed suitably if L_x is of different order; $L_t^{-1}L_t\phi = \phi - \phi(x,0)$ for first-order L_t. To consider the possibilities simultaneously, we write $L_x^{-1}L_x\phi = \phi - \Phi$ where $\Phi = x\alpha(t) + \beta(t)$ in the second-order case, for example, and can be considered known from the specified boundary conditions. (Thus for $\Phi = x\alpha(t) + \beta(t)$ using the boundary conditions, we see that $\beta = g(t)$, $\alpha = 0$.) Applying the inverse operators L_x^{-1} to equation (14.1.3) and L_t^{-1} to equation (14.1.4) and solving for ϕ

$$\phi = g(t) + b^{-1} L_x^{-1} L_t\phi + ab^{-1} L_x^{-1} N\phi$$

$$\phi = f(x) + bL_t^{-1} L_x\phi - aL_t^{-1} N\phi \tag{14.1.5}$$

Adding and dividing by two, and defining

$$\phi_0 = (1/2)\{f(x) + g(t)\}$$

we have

$$\phi = \phi_0 + (1/2)\{bL_t^{-1} L_x + b^{-1} L_x^{-1} L_t\}\phi$$
$$+ (1/2)\{ab^{-1} L_x^{-1} - aL_t^{-1}\}N\phi \tag{14.1.6}$$

Let $\phi = \sum_{n=0}^{\infty} \phi_n$ and $N\phi = \sum_{n=0}^{\infty} A_n$ where the A_n are generated for $\phi^m\phi_x$. For this case, they can be written (1986):

$$A_0 = \phi_0^m (\partial/\partial x)\phi_0$$

$$A_1 = \phi_0^m (\partial/\partial x)\phi_1 + \phi_1^m(\partial/\partial x)\phi_0 \tag{14.1.7}$$

$$A_2 = \phi_0^m (\partial/\partial x)\phi_2 + \phi_1^m(\partial/\partial x)\phi_1 + \phi_2^m(\partial/\partial x)\phi_0$$

Thus, we have

$$\begin{aligned}\phi = \phi_0 &+ (1/2)\{bL_t^{-1} L_x + b^{-1} L_x^{-1} L_t\}\sum_{n=0}^{\infty} \phi_n \\ &+ (1/2)\{ab^{-1} L_x^{-1} - aL_t^{-1}\}\sum_{n=0}^{\infty} A_n\end{aligned} \tag{14.1.8}$$

which leads to

$$\begin{aligned}\phi_1 &= (1/2)\{bL_t^{-1} L_x + b^{-1} L_x^{-1} L_t\}\phi_0 \\ &\quad + (1/2)\{ab^{-1} L_x^{-1} - aL_t^{-1}\} A_0 \\ \phi_2 &= (1/2)\{bL_t^{-1} L_x + b^{-1} L_x^{-1} L_t\}\phi_1 \\ &\quad + (1/2)\{ab^{-1} L_x^{-1} - aL_t^{-1}\}A_1 \\ &\vdots\end{aligned} \tag{14.1.9}$$

Thus for $n \geq \phi$

$$\begin{aligned}\phi_{n+1} &= (1/2)\{bL_t^{-1} L_x + b^{-1} L_x^{-1} L_t\}\phi_n \\ &\quad + (1/2)\{ab^{-1} L_x^{-1} - aL_t^{-1}\}A_n\end{aligned} \tag{14.1.10}$$

allows us to determine all components. The sum $\sum_{n=0}^{\infty} u_n$ is the complete solution and the n-term approximation ϕ_n is the practical solution converging for reasonable n (1986). Thus the necessary components are easily calculated once the initial/boundary conditions have been specified.

For multi-dimensional extension, let's consider the case:

$$\phi_t + a\phi\phi_x = b\nabla^2\phi$$

or letting $a = b = 1$ for convenience,

$$\phi_t = \phi\phi_x = \nabla^2\phi$$

corresponding with the preceding case when $m = 1$ and $n = 2$. (Changing m to 2 or 3, for example, only means the A_n will change accordingly.) We can write this as

$$[L_t - L_x - L_y - L_z]\phi = \phi\phi_x$$

where $L_t = \partial/\partial t$, $L_x = \partial^2/\partial x^2$, $L_y = \partial^2/\partial y^2$, $L_z = \partial^2/\partial z^2$. Solving for each linear operator in turn

$$L_t\phi = [L_x + L_y + L_z]\phi + \phi\phi_x$$

$$L_x\phi = [L_t - L_y - L_z]\phi - \phi\phi_x$$

$$L_y\phi = [L_t - L_z - L_x]\phi - \phi\phi_x$$

$$L_z\phi = [L_t - L_x - L_y]\phi - \phi\phi_x$$

When we consider $L_t^{-1} L_t\phi$, we get $\phi - \alpha_1$ where $\alpha_1 = \phi(x,0)$. Similarly $L_x^{-1} L_x\phi = \phi - \alpha_2 - \alpha_3 x$ where α_2, α_3 are evaluated from the specified conditions. Continuing, we can write

$$\phi = \alpha_1 + L_t^{-1}(L_x + L_y + L_z)\phi + L_t^{-1}\phi\phi_x$$

$$\phi = (\alpha_2 + \alpha_3 x) + L_x^{-1}(L_t - L_y - L_z)\phi - L_x^{-1}\phi\phi_x$$

$$\phi = (\alpha_4 + \alpha_5 y) + L_y^{-1}(L_t - L_z - L_x)\phi - L_y^{-1}\phi\phi_x$$

$$\phi = (\alpha_6 + \alpha_7 z) + L_z^{-1}(L_t - L_x - L_y)\phi - L_z^{-1}\phi\phi_x$$

Define

$$\phi_0 = (1/4)\{\alpha_1 + (\alpha_2 + \alpha_3 x) + (\alpha_4 + \alpha_5 y) + (\alpha_6 + \alpha_7 z)\}$$

$$K = (1/4)\{L_t^{-1}(L_x + L_y + L_z) + L_x^{-1}(L_t - L_y - L_z) + L_y^{-1}(L_t - L_z - L_x) + L_z^{-1}(L_t - L_x - L_y)\}$$

$$G = (1/4)\{L_t^{-1} - L_x^{-1} - L_y^{-1} - L_z^{-1}\}$$

Now

$$\phi_{n+1} = K\,\phi_n + G\,A_n$$

yields all components for $n \geq 0$ in terms of u_0, which is known, and $\phi = \sum_{n=0}^{\infty} \phi_n$ represents the complete solution. We can carry the approximation as far as necessary for a practical computational solution.

14.2 Dissipative Wave Equations

Consider the equation $u_{tt} - u_{xx} + (\partial/\partial t)f(u) = 0$ with $f(u(x,t))$ a continuous bounded function and $(t,x) \in [0,T] \times R$ and let $L_t = \partial^2/\partial t^2$ and $L_x = \partial^2/\partial x^2$ and write

$$L_t u - L_x u = -(\partial/\partial t)f(u(x,t)) \tag{14.2.1}$$

We solve for each linear term thus:

$$\begin{aligned} L_t u &= L_x u - (\partial/\partial t)f(u) \\ L_x u &= L_t u - (\partial/\partial t)f(u) \end{aligned} \tag{14.2.2}$$

Operating with the inverses, we have $L_t^{-1} L_t u = u - \phi_t$ and $L_x^{-1} L_x u = u - \phi_x$, where the homogeneous solutions are evaluated from the given initial boundary conditions. Thus (14.2.2) becomes

$$u = \phi_t + L_t^{-1} L_x u - L_t^{-1} (\partial/\partial t)f(u)$$

$$u = \phi_x + L_x^{-1} L_t u - L_x^{-1} (\partial/\partial t)f(u)$$

Adding and dividing by two

$$u = (1/2)(\phi_t + \phi_x) + (1/2)(L_t^{-1} L_x + L_x^{-1} L_t)u$$
$$- (1/2(L_t^{-1} - L_x^{-1})(\partial/\partial t)f(u)$$

or if

$$K = (1/2)(L_t^{-1} L_x + L_x^{-1} L_t)$$

$$G = -(1/2)(L_t^{-1} - L_x^{-1}) \qquad (14.2.3)$$

$$u_0 = (1/2)(\phi_t + \phi_x)$$

we have

$$u = u_0 + Ku + G(\partial/\partial t)f(u) \qquad (14.2.4)$$

a result also obtained by operating on (14.2.1) with $(L_t^{-1} - L_x^{-1})$.

Let $u = \sum_{n=0}^{\infty} u_n$ with u_0 as defined, and let $f(u) = \sum_{n=0}^{\infty} A_n$, where the A_n are generated for the analytic function $f(u)$. Now

$$u = u_0 + K \sum_{n=0}^{\infty} u_n + G(\partial/\partial t)\sum_{n=0}^{\infty} A_n \qquad (14.2.5)$$

We define

$$u_{n+1} = Ku_n + G(\partial/\partial t)A_n \qquad (14.2.6)$$

for $n \geq 0$ to complete the solution.

References

Adomian, G. *Nonlinear Stochastic Operator Equations*, Academic Press, 1986.

Oguztoreli, M. N., E. S. Shunubi, and K. V. Leung, "Wave Propagation in Dissipative or Dispersive Nonlinear Media: A Numerical Approach," *Applied Math. and Computation,* 6, 4, 1980, 309-334.

CHAPTER 15

The Nonlinear Klein-Gordon Equation

We can consider the equation $u_{tt} - \nabla^2 u = g(u)$ and obtain a solution for study using decomposition (1986). Letting $L_t = \partial^2/\partial t^2$, $L_x = \partial^2/\partial x^2$, $L_y = \partial^2/\partial y^2$, $L_z = \partial^2/\partial z^2$, we rewrite the equation as

$$[L_t - L_x - L_y - L_z]u = g(u)$$

Solving for each linear operator term in turn (1986) we have:

$$L_t u = [L_x + L_y + L_z]u + g(u)$$

$$L_x u = [L_t - L_y - L_z]u - g(u)$$

$$L_y u = [L_t - L_x - L_z]u - g(u)$$

$$L_z u = [L_t - L_x - L_y]u - g(u)$$

Operate on each equation by the appropriate inverse. Thus $L_t^{-1} L_t u = u - \alpha_1 - \alpha_2 t$ where α_1, α_2 are determined from the initial/boundary conditions, e.g., $\alpha_1 = u(0,x,y,z)$, $\alpha_2 = u_t(0,x,y,z)$. We deal the same way with the following equations. Now

$$u = (\alpha_1 + \alpha_2 t) + L_t^{-1}[L_x + L_y + L_z]u + L_t^{-1} g(u)$$

$$u = (\alpha_3 + \alpha_4 x) + L_x^{-1}[L_t - L_y - L_z]u - L_x^{-1} g(u)$$

$$u = (\alpha_5 + \alpha_6 y) + L_y^{-1}[L_t - L_x - L_z]u - L_y^{-1} g(u)$$

$$u = (\alpha_7 + \alpha_8 z) + L_z^{-1}[L_t - L_x - L_y]u - L_z^{-1} g(u)$$

If we have initial conditions specified, the simplest procedure is to add the four equations and divide by four to get an equation for u

$$u = (1/4)\{(\alpha_1 + \alpha_2 t) + (\alpha_3 + \alpha_4 x) + (\alpha_5 + \alpha_6 y) + (\alpha_7 + \alpha_8 z)\}$$
$$+ (1/4)\{L_t^{-1} [L_x + L_y + L_z] + L_x^{-1} [L_t - L_y - L_z]$$
$$+ L_y^{-1} [L_t - L_x - L_z] + L_z^{-1} [L_t - L_x - L_z]\}u$$
$$+ (1/4)[L_t^{-1} - L_x^{-1} - L_y^{-1} - L_z^{-1}] g(u)$$

Define

$$u_0 = (1/4)\{(\alpha_1 + \alpha_2 t) + (\alpha_3 + \alpha_4 x) + (\alpha_5 + \alpha_6 y) + (\alpha_7 + \alpha_8 z)\}$$
$$K = (1/4)\{L_t^{-1} (L_x + L_y + L_z] + L_x^{-1} \cdot (L_y - L_z - L_t)$$
$$+ L_y^{-1} (L_z - L_x - L_t) + L_z^{-1} (L_x - L_y - L_t)\}$$
$$G = (1/4)\{L_t^{-1} - L_x^{-1} - L_y^{-1} - L_z^{-1}\}$$

and $u = \sum_{n=0}^{\infty} u_n$ and $g(u) = \sum_{n=0}^{\infty} A_n$ where the A_n are evaluated specifically for $g(u)$.

Now since u_0 is known, all further components are calculable by

$$u_{n+1} = K u_n + G A_n$$

for $n \geq 0$. The solution avoids linearization and other common approximations.

When the specified conditions are boundary values, we evaluate u_0 for each of the four equations using the appropriate boundary values, then determine each u_1 for each of the four equations from the previous u_0 in the same equation getting constants of integration which are again found from the boundary conditions. Proceeding to some u_{n-1}, we write $\phi_n = \sum_{i=0}^{\infty} u_i$.

Consider a simple linear example: $u_{xx} - u_{yy} = 0$ on $0 \leq x \leq \pi/2$, $0 \leq y \leq \pi/2$ given conditions:

$$u(0,y) = 0 \qquad u(\pi/2,y) = \sin y$$

$$u(x,0) = 0 \qquad u(x,\pi/2) = \sin x$$

Let $L_x = \partial^2/\partial x^2$ and $L_y = \partial^2/\partial y^2$ and write the above equation as $L_x u = L_y u$.

As usual in the decomposition method, we solve for each linear operator term, $L_x u$ and $L_y u$, in turn and then apply the appropriate inverse to each.

$$L_x^{-1} L_x u = u - c_1 k_1(y) - c_2 k_2(y) x = L_x^{-1} L_y u$$

$$L_y^{-1} L_y u = u - c_3 k_3(x) - c_4 k_4(x) y = L_y^{-1} L_x u$$

or

$$u = c_1 k_1(y) + c_2 k_2(y) x + L_x^{-1} L_y u \tag{15.1}$$

$$u = c_3 k_3(y) + c_4 k_4(x) y + L_y^{-1} L_x u \tag{15.2}$$

Define $\phi_x = c_1 k_1(y) + c_2 k_2(y) x$ and $\phi_y = c_3 k_3(y) + c_4 k_4(x) y$ to rewrite (15.1) and (15.2) as

$$u = \phi_x + L_x^{-1} L_y u \tag{15.3}$$

$$u = \phi_y + L_y^{-1} L_x u \tag{15.4}$$

One-term approximants to the solution u are $u_0 = \phi_x$ in (15.3) and $u_0 = \phi_y$ in (15.4). Two-term approximants are $u_0 + u_1$ where $u_1 = L_x^{-1} L_y u_0$ in (15.3) and $L_y^{-1} L_x u_0$ in (15.4) etc. Thus $u_{n+1} = L_x^{-1} L_y u_n$ in (15.3) and $L_y^{-1} L_x u_n$ in (15.4) for $n \geq 0$.

For the x conditions $u(0,y) = 0$ and $u(\pi/2,y) = \sin y$ applied to the one-term approximant $u_0 = c_1 k_1(y) + c_2 k_2(y) x$, we have

$$c_1 k_1(y) = 0$$

$$c_2 k_2(y) \pi/2 = \sin y$$

or $c_2 = 2/\pi$ and $k_2(y) = \sin y$.

For the y conditions $u(x,0) = 0$ and $u(x, \pi/2) = \sin x$ applied to $u_0 = c_3 k_3(x) + c_4 k_4(x) y$, we get

$$c_3 k_3(x) = 0$$

$$c_4 k_4(x) \pi/2 = \sin x$$

Thus $c_4 = 2/\pi$ and $k_4(x) = \sin x$. If a one-term approximant were sufficient, the solution would be

$$\Phi_1 = (1/2)\{(2/\pi) x \sin y + (2/\pi) y \sin x\}$$

The next terms for (15.3) and (15.4) respectively are

$$u_1 = L_x^{-1} L_y u_0 = L_x^{-1} L_y [c_2 x \sin y]$$

$$u_1 = L_y^{-1} L_x u_0 = L_y^{-1} L_x [c_4 y \sin x]$$

We continue to obtain $u_2, u_3, \ldots$. Clearly, for any n,

$$u_n = (L_x^{-1} L_y)^n u_0 = c_2 (\sin y)(-1)^n x^{2n+1}/(2n+1)!$$

$$u_n = (L_y^{-1} L_x)^n u_0 = c_4 (\sin x)(-1)^n y^{2n+1}/(2n+1)!$$

Letting Φ_m represent the m-term approximant, we have for two cases:

$$\Phi_m = c_2 \sin y \sum_{n=0}^{m-1} (-1)^n x^{2n+1}/(2n+1)! \quad (15.5)$$

$$\Phi_m = c_4 \sin x \sum_{n=0}^{m-1} (-1)^n y^{2n+1}/(2n+1)! \quad (15.6)$$

We can now apply the conditions $\Phi_m(\pi/2, y) = \sin y$ for (15.5); thus

$$c_1 k_1(y) = 0$$

$$c_2 \sin y \sum_{n=0}^{m-1} (\pi/2)^{2n+1}/(2n+1)! = \sin y$$

$$c_2 = \frac{1}{\sum_{n=0}^{m-1} (-1)^n(\pi/2)^{2n+1}/(2n+1)!}$$

As $m \to \infty$, we get $\sin \pi/2$ so that $c_2 \to 1$. The sum in (15.5) approaches $\sin x$ in the limit.

Now applying the conditions $\Phi_m(x,0) = 0$ and $\Phi_m(x,\pi/2) = \sin x$, we have

$$c_3k_3(x) = 0$$

$$c_4 \sin x \sum_{n=0}^{m-1} (-1)^n(\pi/2)^{2n+1}/(2n+1)! = \sin x$$

$$c_4 = \frac{1}{\sum_{n=0}^{m-1} (-1)^n(\pi/2)^{2n+1}/(2n+1)!}$$

Again as $m \to \infty$. $c_4 \to 1$ and the sum in (15.6) becomes $\sin y$. We can now write the exact solution

$$u = (1/2)\{\sin y \sin x + \sin x \sin y\}$$

or

$$u = \sin y \sin x$$

since for this case, the series is summed.

In some cases, u_0 may vanish in one equation in which case the remaining equations are sufficient to provide the solution even more simply. Consider $u_{xx} = u_{tt}$ on $0 \leq x \leq \pi$ with $t \geq 0$ specifying the conditions

$u(x,0) = \sin x$

$u_t(x,0) = 0$

$u(0,t) = 0$

$u(\pi,t) = 0$

By decomposition we have

$$u = c_1k_1(t) + c_2k_2(t)\,x + L_x^{-1} L_t u \tag{15.7}$$

$$u = c_3k_3(t) + c_4k_4(t)\,t + L_t^{-1} L_x u \tag{15.8}$$

Thus the one-term approximant $\phi_1 = u_0$ in (15.7) is $u_0 = c_1k_1(t) + c_2k_2(t)x$. Satisfying conditions on x, we find $c_1k_1(t) = 0$ and $c_2k_2(t)\pi = 0$, so $u_0 = 0$. Thus (15.7) does not contribute and is not used. From (15.8), $u_0 = c_3k_3(x) + c_4k_4(x)t$ and, applying the t conditions, $c_3k_3(x) = \sin x$ and $c_4k_4(x)t = 0$. Hence

$u_0 = \sin x$

$u_1 = L_t^{-1} L_x u_0 = L_t^{-1} L_x \sin x$

$u_1 = (-t^2/2) \sin x$

$u_2 = (t^4/4!) \sin x$

.

.

.

or $u = (1 - t^2/2! + t^4! - \ldots) \sin x$ and finally

$u = \sin x \cos t$

A similar example occurs in the diffusion equation $u_t - u_{xx} = 0$ with the conditions $u(x,0) = \sin x$, $u(0,t) = u(\pi,t) = 0$. From $L_t u = L_{xx} u$, with the application of the inverse L_t^{-1}, we get

$$u = u_0 + L_t^{-1} L_{xx} u$$

with $u_0 = u(x,0) = \sin x$. Then

$$u_1 = L_t^{-1} L_{xx} u_0 = L_t^{-1} L_{xx} \sin x = -t \sin x$$

etc., for $u_{n+1} = L_t^{-1} L_{xx} u_n$ for $n \geq 0$ to yield

$$u = e^{-t} \sin x$$

We see the equation using L_{xx}^{-1} does not contribute because the $u_0 = A + Bx \equiv 0$ when we satisfy $u(0,t) = u(\pi,t) = 0$.

It is easily shown that the solutions obtained do satisfy the given ordinary or partial differential equations and specified conditions. Since the method has been applied successfully to a wide class of examples of linear and nonlinear ordinary, partial, integro-differential, and integral equations, as well as systems of equations, it is impossible to make a general statement about the number of terms required.

We have chosen a linear equation as a convenient example to show how the method works in practice. The fact that the example is linear is of no relevance. One can equally well think of the procedure as expanding both the solution process u and any nonlinear function $g(u)$ in the A_n polynomials (instead of saying $u = \sum_{n=0}^{\infty} u_n$ and $g(u) = u = \sum_{n=0}^{\infty} A_n$ where the A_n are generated specifically for $g(u)$). Since for any linear or nonlinear functions, $A_0 = A_0(u_0)$, $A_1 = A_1(u_0,u_1),\ldots, A_n = A_n(u_0,u_1,\ldots u_n)$, the nonlinear problem can be solved exactly as in the case of the linear problem.

The solution can now be determined for the given conditions which are separable - the α's are functions, e.g., at $t = 0$, $u_0 = \alpha_1(x,y,z)$. Curved surfaces are not considered here. The method discussed (1986) has now worked remarkably well in a large number of physical problems and may yield insights not otherwise possible in nonlinear problems. Our objective is to obtain unique solutions for various possible specified conditions for real physical problems; however much remains to be done in carefully defining the ranges of applicability and conditions for existence of solutions.

Reference

Adomian, G., *Nonlinear Stochastic Operator Equations* , Academic Press, 1986.

Chapter 16

Analysis of Model Equations of Gas Dynamics

In this chapter we consider the application of our decomposition method as a possible approach to the analysis of equations modeling gas dynamics with given initial data. Consider the model equation (1986):

$$u_t + (1/2)(u^2)_x = u(1 - u)$$

Let $L = \partial/\partial t$ and write

$$Lu = u - u^2 - (1/2)(u^2)_x$$

Defining L^{-1} as the definite integral from 0 to t and operating on both sides with L^{-1}, we obtain

$$u = u(x,0) + L^{-1}u - L^{-1}u^2 - (1/2)\, L^{-1}(u^2)_x$$

Using decomposition, let $u = \sum_{n=0}^{\infty} u_n$ and for initial data, assume $u(x,0) = g(x) = b(1 - e^{-x})$, where $b > 0$ (1987). Now expand u^2 and $(u^2)_x$ in the A_n polynomials . For u^2 we have:

$$A_0 = u_0^2$$

$$A_1 = 2u_0u_1$$

$$A_2 = u^2 + 2u_0u_2$$

$$\vdots$$

and for $(u^2)_x$, we have

$$A_0 = u_{0x}^2$$

$$A_1 = 2u_{0x}\, u_{1x}$$

$$A_2 = u_{1x}^2\ 2u_{0x}u_{2x}$$

$\vdots$

We can now compute the components of u :

$$u_0 = g(x) = b(1 - e^{-x})$$

$$u_1 = L^{-1}u_0 - L^{-1}u_0^2 - (1/2)L^{-1}(u_0)_x^2$$

$$u_2 = L^{-1}u_1 - L^{-1}(2u_0u_1) - (1/2)L^{-1}2(u_0)_x(u_1)_x$$

$$u_3 = L^{-1}u_2 - L^{-1}(u_1^2 + 2u_0u_2) - (1/2)L^{-1}[(u_1)_x^2 + 2(u_0)_x(u_2)_x]$$

$\vdots$

Consequently, we have

$$u_0 = b(1 - e^{-x})$$

$$u_1 = bt(1 - e^{-x}) - b^2t(1 - e^{-x})^2 - (1/2)b^2te^{-2x}$$

$$u_2 = (bt^2/2)(1 - e^{-x}) - (b^2t^2/2)(1 - e^{-x})^2$$

$$- (b^2t^2/4)e^{-2x} - b^2t^2(1 - e^{-x})^2$$

$$- b^3t^2(1 - e^{-x})^3 - (b^3t^2/2)e^{-2x}(1 - e^{-2x})$$

$\vdots$

so that

$$u = b(1 - e^{-x}) + bt(1 - e^{-x}) - b^2t(1 - e^{-x})^2$$

$$- (1/2)b^2t\, e^{-2x} + (bt^2/2\ (1 - e^{-x})$$

$$+ (b^2t^2/2)(1 - e^{-x})^2 - (b^2t^2/4)/e^{-2x}$$

$$- b^2t^2(1 - e^{-x})^2 - b^3t^2(1 - e^{-x})^3$$

$$- (b^3t^2/2)\, e^{-2x}\, (1 - e^{-2x}) + \ldots$$

Other forms of gas dynamic model equations are solvable in the same manner. Different nonlinear terms simply result in different A_n polynomials. Forcing terms will modify the u_0 term and the resulting series. Possible modifications to the model equations are numerous, and we will not attempt to deal with all the possibilties here. General procedures have been fully discussed in (1986) and should be applicable with possible decreased demands on computer resources. The primary problem is realistic modeling rather than tractability of equations.

References

Adomian, G., *Nonlinear Stochastic Operator Equations* , Academic Press, 1986.

Bellomo, N. and E. Longo, "On the Discretization of the Boltzmann Equation for Binary Gas Mixtures: Theory and Application in Molecular Fluid Dynamics," *L'Aerotechnica. J. Italian Assoc. Aeron. Astron.* , 3, 1983, 146-153.

Salas, M. D., S. Abarbanel, and D. Gottlieb, "Multiple Steady States for Characteristic Initial-Value Problems," ICASE Report, 1987.

CHAPTER 17

A New Approach to the Efinger Model for a Nonlinear Quantum Theory for Gravitating Particles

H. Efinger has proposed a self-consistent (deterministic, nonrelativistic) model for quantum theory and gravitation (1973;1980, 1984). We propose here to consider this model.

One dimensional case: In this case, Efinger's equations are

$$L_1X = \kappa|Y|^2 \tag{17.1}$$

$$L_2Y = \kappa XY \tag{17.2}$$

with κ a coupling constant (κ^2 is proportional to the gravitational constant) and where L_1, L_2 are differential operators and appropriate initial and boundary conditions are given. From (17.1)

$$X = \phi_x + L_1^{-1} \kappa|Y|^2 \tag{17.3}$$

where $L_1\phi_x = 0$.

By decomposition, we write $X = \sum_{n=0}^{\infty} X_n$ and $Y = \sum_{n=0}^{\infty} Y_n$

$$\begin{aligned} \sum_{n=0}^{\infty} X_n &= \phi_x + L_1^{-1} \kappa|Y|^2 \\ \sum_{n=0}^{\infty} Y_n &= \phi_y + L_2^{-1} \kappa XY \end{aligned} \tag{17.4}$$

We identify $X_0 = \phi_x$ and $Y_0 = \phi_y$ where ϕ_x, ϕ_y include the initial/boundary conditions and are easily evaluated when these conditions and the operators L_1, L_2 are specified. These operators can be decomposed into an easily invertible part (the highest ordered derivative) and a remainder operator avoiding a Green's function which would lead to a difficult integral. Other components are now easily calculated.

We replace $|Y^2|$ by $\sum_{n=0}^{\infty} A_n(|Y|^2)$, i.e., the A_n polynomials generated for $|Y|^2$ and replace XY by $\sum_{n=0}^{\infty} A_n(XY)$ generated for XY. Now

$$X_1 = L_1^{-1} \kappa A_0(|Y|^2)$$

$$Y_1 = L_2^{-1} \kappa A_0 (XY)$$

Since $A_0(|Y|^2) = |Y_0|^2$ and $A_0(XY) = X_0Y_0$, we have

$$X_1 = L_1^{-1} \kappa |Y_0|^2$$

$$Y_1 = L_2^{-1} \kappa X_0Y_0$$

Thus X_1, Y_1 are calculable interms of X_0, Y_0. For $A_1(|Y|^2)$ we have $2 Y_0Y_1$ and for $A_1(XY) = X_0Y_1 + X_1Y_0$ hence:

$$X_2 = L_1^{-1} \kappa(2Y_0Y_1)$$

$$Y_2 = L_2^{-1} \kappa (X_0Y_1 + Y_1Y_0)$$

So again, X_2, Y_2 are seen to be calculable. The A_n can now be computed; thus all components X_n, Y_n for $n \geq 1$ are easily calculated. n-term approximations $\sum_{i=0}^{\infty} X_i$ and $\sum_{i=0}^{n-1} Y_i$ serve as the solutions, and as $n \to \infty$ we get $\sum_{n=0}^{\infty} X_n = X$ and $\sum_{n=0}^{\infty} Y_n = Y$. The procedure is practical and easily generalizable to three space dimensions and time. First suppose the operators L_1, L_2 are given by

$$L_1 = d^2/dt^2 - 1$$

$$L_2 = d^2/dt^2 + \varepsilon$$

where ε is an eigenvalue parameter. If we let $L = d^2/dt^2$, equations (17.1) and (17.2) become

$$\begin{aligned} LX &= \kappa |Y|^2 + X \\ LY &= \kappa XY - \varepsilon Y \end{aligned} \qquad (17.5)$$

Thus

$$\sum_{n=0}^{\infty} X_n = \phi_x + L^{-1}\,\kappa\,|Y|^2 + L^{-1}\sum_{n=0}^{\infty} X_n$$

$$\sum_{n=0}^{\infty} Y_n = \phi_y + L^{-1}\,\kappa\,XY - L^{-1}\,\varepsilon\sum_{n=0}^{\infty} Y_n$$

Again X_0, Y_0 are known and following pairs X_n, Y_n for $n > 1$ are determined from

$$X_1 = L^{-1}\,\kappa\,A_0(|Y|^2) + L^{-1}\,X_0$$

$$Y_1 = L^{-1}\,\kappa\,A_0(XY) - L^{-1}\,\varepsilon\,Y_0$$

etc. to determine components as needed. Note particularly that the integrations are simple two-fold integrations without a kernel.

Three Dimensions: Suppose we now write $L_1 = \nabla^2 - 1$ and $L_2 = \nabla^2 + \varepsilon$. We write $L_x = \partial^2/\partial x^2$, $L_y = \partial^2/\partial y^2$, $L_z = \partial^2/\partial z^2$; hence $L_1 = L_x + L_y + L_z - 1$ and $L_2 = L_x + L_y + L_z + \varepsilon$. Now we have

$$(L_x + L_y + L_z)X = \kappa\,|Y|^2 + X$$

$$(L_x + L_y + L_z)Y = \kappa\,XY - \varepsilon Y$$

which we must solve for each linear operator in turn thus:

$$L_xX = \kappa|Y|^2 + X - L_yX - L_zX$$

$$L_yX = \kappa|Y|^2 + X - L_xX - L_zX$$

$$L_zX = \kappa|Y|^2 + X - L_xX - L_yX$$

and

$$L_xY = \kappa XY - \varepsilon Y - L_yY - L_zY$$

$$L_yY = \kappa XY - \varepsilon Y - L_xY - L_zY$$

$$L_zY = \kappa XY - \varepsilon Y - L_xY - L_yY$$

Inverting as before,

$$X = \phi_x + L_x^{-1} \kappa |Y|^2 + L_x^{-1} X - L_x^{-1} L_y X - L_x^{-1} L_z X$$

$$X = \phi_y + L_y^{-1} \kappa |Y|^2 + L_y^{-1} X - L_y^{-1} L_x X - L_y^{-1} L_z X$$

$$X = \phi_z + L_z^{-1} \kappa |Y|^2 + L_z^{-1} X - L_z^{-1} L_x X - L_z^{-1} L_y X$$

$$Y = \theta_x + L_x^{-1} \kappa XY - L_x^{-1} \varepsilon Y - L_x^{-1} L_y Y - L_x^{-1} L_z Y$$

$$Y = \theta_y + L_y^{-1} \kappa XY - L_y^{-1} \varepsilon Y - L_y^{-1} L_x Y - L_y^{-1} L_z Y$$

$$Y = \theta_z + L_z^{-1} \kappa XY - L_z^{-1} \varepsilon Y - L_z^{-1} L_x Y - L_z^{-1} L_y Y$$

The equations for X are added and divided by three, and the equations for Y are added and divided by three. The result is

$$X = (1/3)(\phi_x + \phi_y + \phi_z) + (1/3)(L_x^{-1} + L_y^{-1} + L_z^{-1})\kappa |Y|^2$$
$$+ (1/3)(L_x^{-1} + L_y^{-1} + L_z^{-1})X$$
$$- (1/3)(L_x^{-1} L_y + L_y^{-1} L_x + L_z^{-1} L_x + L_x^{-1} L_z + L_y^{-1} L_z + L_z^{-1} L_y)X$$

$$Y = (1/3)(\theta_x + \theta_y + \theta_z) + (1/3)(L_x^{-1} + L_y^{-1} + L_z^{-1})\kappa XY$$
$$+ (1/3)(L_x^{-1} + L_y^{-1} + L_z^{-1})\varepsilon Y$$
$$- (1/3)(L_x^{-1} L_y + L_y^{-1} L_x + L_z^{-1} L_x + L_x^{-1} L_z + L_y^{-1} L_z + L_z^{-1} L_y)Y$$

Identify

$$X_0 = (1/3)(\phi_x + \phi_y + \phi_z)$$

$$Y_0 = (1/3)(\theta_x + \theta_y + \theta_z)$$

and substitute again the A_n polynomials for the nonlinear terms and the decompositions $\sum_{n=0}^{\infty} X_n$ for X and $\sum_{n=0}^{\infty} Y_n$ for Y. Now

$$X_{n+1} = (1/3)(L_x^{-1} + L_y^{-1} + L_z^{-1})\kappa A_n(|Y|^2)$$
$$+ (1/3)(L_x^{-1} + L_y^{-1} + L_z^{-1})X_n$$
$$- (1/3)(L_x^{-1} L_y + L_y^{-1} L_x + L_z^{-1} L_x + L_x^{-1} L_z + L_y^{-1} L_z + L_z^{-1} L_y)X_n$$
$$Y_{n+1} = (1/3)(L_x^{-1} + L_y^{-1} + L_z^{-1})\kappa A_n(XY)$$
$$- (1/3)(L_x^{-1} + L_y^{-1} + L_z^{-1})\varepsilon Y_n$$
$$- (1/3)(L_x^{-1} L_y + L_y^{-1} L_x + L_z^{-1} L_x + L_x^{-1} L_z + L_y^{-1} L_z + L_z^{-1} L_y)Y_n$$

determine all components for $n \geq 0$.

Generalizations to four dimensions (spacetime), stochastic formulation, more complex operators or nonlinearities, and boundary conditions which may be nonlinear, stochastic, or even coupled now appear to be within the scope of the methodology (1986) which means our attention can now be devoted to understanding the physics and the effects or significance of a stochastic generalization.

References

Adomian, G., *Nonlinear Stochastic Operator Equations* , Academic Press, 1986.

Efinger, H., "On Certain Nonlinear Equations Arising in Self-Consistent Theories," Univ. of Ga. Ctr. for App. Math. Report CAM 26, *Il Nuovo Cimento* , 59B, 1980, 314.

Efinger, H., "A Model for Gravitating Particles in Quantum Theory," *Act. Phys. Austr.* , 37, 1973, 343.

Efinger, H., "On the Theory of Certain Nonlinear Schrodinger Equations with Nonlocal Interaction, " *Il Nuovo Cimento* , 80B, 1984, 260.

CHAPTER 18

The Navier-Stokes Equations

The basic dynamical equations of fluid mechanics are represented by the Navier-Stokes equations. These are nonlinear partial differential equations describing the phenomena of fluid flow obtained by applying Newton's second law of motion to fluids. They apply to problems as diverse as flow of air around aircraft, airflow in ramjet or scramjet engines, and the blood circulation in humans (with modifications considering elastic artery walls, etc.). A complete solution under hypersonic flow conditions could lead to aircraft velocities of Mach 10 to Mach 20 and achieve single-stage-to-orbital flight. These equations have defied analytical solution for a century, and analytical approaches have been replaced by numerical methods which discretize the problem and lead to severe problems of computational time on supercomputers.

Since it is now possible to solve nonlinear, and nonlinear stochastic, partial differential equation systems without a need for linearization or assumptions of "weak" nonlinearity, "small" fluctuations, and discretization, our objective will be to seek continuous verifiable analytic solutions without the massive printouts and restrictive assumptions which necessarily change the physical problem into a mathematically tractable and different problem which will not yield the same solution. Such restrictive methods are of little use in cases involving strongly nonlinear, strongly stochastic phenomena such as turbulent flow. What we need is a solution retaining these intrinsic features and we must work towards this objective.

The hypersonic case offers fascinating vistas for the future but is not just a matter of faster computers or of application of decomposition. Additional thermodynamic and chemical effects will modify the equations we are considering and make them even more complicated. Consideration will have to await proper modeling. We consider the well-known Navier-Stokes equations here by decomposition in the hope it will provide insights and a new approach with possible advantages.

Analysis: Consider a Cartesian system of x, y, z axes with unit vectors $\hat{i}, \hat{j}, \hat{k}$ with $\hat{k}$ upwards, i towards the East and $\hat{j}$

towards the North fixed on the surface of the ocean. Represent velocity by $\overline{u}$ with components $u(x,y,z,t)$, $v(x,y,z,t)$, $w(x,y,z,t)$. In general, pressure p, density ρ, and viscosity μ can be allowed to be functions of x, y, z, t. They may also depend on ω where $\omega \varepsilon (\Omega, F, \mu)$, a probability space. The Navier-Stokes equation, including an induced force $\overline{F}$ and the Coriolis force due to the earth's rotation Ω_e is written in the form:

$$\begin{aligned} &\rho D\overline{u}/Dt + 2\rho(\Omega_e \times \overline{u}) + \nabla p + kg\rho \\ &\quad = \overline{F} + \mu\nabla^2\overline{u} + (\mu/3)\nabla(\nabla \cdot \overline{u}) \end{aligned} \tag{18.1}$$

where the hydrodynamic derivative D/Dt is given by

$$D/Dt = \partial/\partial t + u\, \partial/\partial x + v\, \partial/\partial y + w\, \partial/\partial z$$

and the continuity condition is written

$$D\rho/Dt + \rho\, \nabla \cdot \overline{u} = 0 \tag{18.2}$$

Equivalently

$$\begin{aligned} &\rho\, \partial\overline{u}/\partial t + \rho(\overline{u} \cdot \nabla)\overline{u} + 2\rho(\Omega \times \overline{u}) + \nabla p + kg\rho \\ &\quad = \overline{F} + \mu\nabla^2\overline{u} + (\mu/3)\nabla(\nabla \cdot \overline{u}) \end{aligned} \tag{18.3}$$

Substituting the relationships

$$(\Omega \times \overline{u}) = \begin{vmatrix} \hat{i} & \hat{j} & \hat{k} \\ \alpha & \beta & \Upsilon \\ u & v & w \end{vmatrix} = \hat{i}(\beta w - \Upsilon v) + \hat{j}(\Upsilon u - \alpha w) + \hat{k}(\alpha v - \beta u)$$

$$\nabla p = \hat{i}\, \partial p/\partial x + \hat{j}\, \partial p/\partial y + \hat{k}\, \partial p/\partial z$$

$$\nabla^2\overline{u} = \hat{i}\nabla^2 u + \hat{j}\nabla^2 v + \hat{k}\nabla^2 w$$

$$\begin{aligned} \nabla(\nabla \cdot \overline{u}) &= \nabla[(1/\rho)(D\rho/Dt)] = \nabla[(1/\rho)\{\partial\rho/\partial t + u\, \partial\rho/\partial x \\ &\quad + v\, \partial\rho/\partial y + w\, \partial\rho/\partial z\}] \end{aligned}$$

$$(\overline{u} \cdot \nabla)\overline{u} = \hat{i}(u\ \partial u/\partial x + v\ \partial u/\partial y + w\ \partial u/\partial z)$$

$$+ \hat{j}\,(u\ \partial v/\partial x + v\ \partial v/\partial y + w\ \partial v/\partial z$$

$$+ \hat{k}(u\ \partial w/\partial x + v\ \partial w/\partial y + w\ \partial w/\partial z)$$

$$\nabla^2\overline{u} = \hat{i}\nabla^2 u + \hat{j}\nabla^2 v + \hat{k}\nabla^2 w$$

and writing the Navier-Stokes equation in component form yields a system of coupled equations.

$$\rho\ \partial u/\partial t + \rho[u\ \partial u/\partial x + v\ \partial u/\partial y + w\ \partial u/\partial z]$$

$$+ 2\rho[\beta w - \Upsilon v] + \partial p/\partial x$$

$$= F_1 + \mu[\partial^2 u/\partial x^2 + \partial^2 u/\partial y^2 + \partial^2 u/\partial z^2]$$

$$+ (\mu/3)[\partial/\partial x(1/\rho)(\partial\rho/\partial t)$$

$$+ (\partial/\partial x)(1/\rho)(u\ \partial\rho/\partial x) + (\partial/\partial x)(1/\rho)(v\ \partial\rho/\partial y)$$

$$+ (\partial/\partial x)(1/\rho)(w\ \partial\rho/\partial z)]$$

$$\rho\ \partial v/\partial t + \rho[u\ \partial v/\partial x + v\ \partial v/\partial y + w\ \partial v/\partial z]$$

$$+ 2\rho[\Upsilon u - \alpha w] + \partial p/\partial y$$

$$= F_2 + \mu[\partial^2 v/\partial x^2 + \partial^2 v/\partial y^2 + \partial^2 v/\partial z^2] \qquad (18.4)$$

$$+ (\mu/3)[\partial/\partial y(1/\rho)(\partial\rho/\partial t)$$

$$+ (\partial/\partial y)(1/\rho)(u\ \partial\rho/\partial x) + (\partial/\partial y)(1/\rho)(v\ \partial\rho/\partial y)$$

$$+ (\partial/\partial y)(1/\rho)(w\ \partial\rho/\partial z)]$$

$$\rho\ \partial w/\partial t + \rho[u\ \partial w \partial x + v\ \partial w/\partial y + w\ \partial w/\partial z]$$

$$+ 2\rho[\alpha v - \beta u] + \partial p/\partial z + g\rho$$

$$= F_3 + \mu[\partial^2 w/\partial x^2 + \partial^2 w/\partial y^2 + \partial^2 w/\partial z^2]$$

$$+ (\mu/3)[(\partial/\partial z)(1/\rho)(\partial\rho/\partial t) + (\partial/\partial z)(1/\rho)(u\ \partial\rho/\partial x)$$

$$+ (\partial/\partial z)(1/\rho)(v\ \partial\rho/\partial y) + (\partial/\partial z)(1/\rho)(w\ \partial\rho/\partial z)]$$

These equations can be written as:

$$Lu + R_1(u,v,w) + N_1(u,v,w) = g_1$$

$$Lv + R_2(u,v,w) + N_2(u,v,w) = g_2 \qquad (18.5)$$

$$Lw + R_3(u,v,w) + N_3(u,v,w) = g_3$$

where we define

$$L = \rho\ \partial/\partial t + \mu\ \partial^2/\partial x^2 + \mu\ \partial^2/\partial y^2 + \mu\ \partial^2/\partial z^2 = L_t + L_x + L_y + L_z$$

$$R_1 = 2\rho(\beta w - \Upsilon v)$$

$$R_2 = 2\rho(\Upsilon u - \alpha w)$$

$$R_3 = 2\rho(\alpha v - \beta u)$$

$$N_1(u,v,w) = \rho[u\ \partial u/\partial x + v\ \partial u/\partial y + w\ \partial u/\partial z]$$

$$- (\mu/3)[(\partial/\partial x)(1/\rho)(u\ \partial\rho/\partial x) + (\partial/\partial x)(1/\rho)(v\ \partial\rho/\partial y)$$

$$+ (\partial/\partial x)(1/\rho)(w\ \partial\rho/\partial z)]$$

$$N_2(u,v,w) = \rho[u\ \partial v/\partial x + v\ \partial v/\partial y + w\ \partial v/\partial z]$$

$$- (\mu/3)[(\partial/\partial y)(1/\rho)(u\ \partial\rho/\partial x) + (\partial/\partial y)(1/\rho)(v\ \partial\rho/\partial y)$$

$$+ (\partial/\partial y)(1/\rho)(w\ \partial\rho/\partial z)]$$

$$N_3(u,v,w) = \rho[u\ \partial w/\partial x + v\ \partial\rho/\partial y + w\ \partial\rho/\partial z]$$

$$- (\mu/3)[(\partial/\partial z)(1/\rho)(u\ \partial\rho/\partial x) + (\partial/\partial z)(1/\rho)(v\ \partial\rho/\partial y)$$

$$+ (\partial/\partial z)(1/\rho)(w\ \partial\rho/\partial z)]$$

$$g_1 = F_1 - \partial\rho/\partial x + (\mu/3)(\partial/\partial x)(1/\rho)(\partial\rho/\partial t)$$

$$g_2 = F_2 - \partial p/\partial y + (\mu/3)(\partial/\partial y)(1/\rho)(\partial \rho/\partial t)$$

$$g_3 = F_3 - \partial p/\partial z + (\mu/3)(\partial/\partial z)(1/\rho)(\partial \rho/\partial t) + g\rho$$

The system of (18.5) will be considered for $x, y, z, t \geq 0$ and initial/boundary conditions specified for particular applications. The nonlinear terms require the use of the A_n polynomials, and we write $u = \sum_{n=0}^{\infty} u_n$, $v = \sum_{n=0}^{\infty} v_n$, $w = \sum_{n=0}^{\infty} w_n$. The inverse of L is found most easily using the form $L = L_t + L_x + L_y + L_z$, solving for each linear operator in turn and applying the appropriate inverse which brings in the initial/boundary conditions. The resulting four equations for each velocity component u, v, w are added and divided by four to yield an equation for u, one for v, and one for w. The (u_0, v_0, w_0) terms of the decompositions are obtained from the initial value (or boundary terms) and the appropriate inverse applied to the forcing function. We evaluate u_0 for each equation (before the addition), proceeding to $u_1, u_2, \ldots,$ and then adding to get ϕ_n. Now that the (u_0, v_0, w_0) triad is known, we can obtain the (u_1, v_1, w_1) triad in terms of (u_0, v_0, w_0). Similarly (u_2, v_2, w_2) is obtained in terms of (u_1, v_1, w_1), etc., up to (u_n, v_n, w_n) in terms of $(u_{n-1}, v_{n-1}, w_{n-1})$ for $n \geq 1$ to some sufficient n for a solution to the desired accuracy.

We defer discussion of the stochastic case until later in the chapter but a comment on a further advantage - and a very important one - of the decomposition method arises in the stochastic case. The generally used methods for stochastic equations require perturbative or hierarchy methods - the latter involving averagings and "closure approximations". The necessity for these unphysical closure approximations is removed in the decomposition method entirely because of the natural statistical separability of the procedure (1986). As the author has pointed out, hierarchy methods are no more than a perturbation method, and we avoid such methods. Because the solution of the full equations without the customary restrictions and simplifying assumptions is quite complicated, let us begin with simpler but helpful models.

1) Solution of a simple form:

$$\partial \overline{u}/\partial t + \overline{\Omega} \times \overline{u} = \overline{F}(\overline{x}, t)$$

where $\overline{F} = (F_1,F_2,F_3)$, $\Omega = (\alpha,\beta,\omega)$, $\overline{u} = (u,v,w)$. Thus

$$\overline{\Omega} \times \overline{u} = \begin{vmatrix} \hat{i} & \hat{j} & \hat{k} \\ \alpha & \beta & \Upsilon \\ u & v & w \end{vmatrix}$$

Thus, in component form,

$$\partial u/\partial t + \beta w - \Upsilon v = F_1$$

$$\partial v/\partial t + \Upsilon u - \alpha w = F_2$$

$$\partial w/\partial t + \alpha v - \beta u = F_3$$

and we see that the curl operator generates a set of coupled equations. Define $L_t = \partial/\partial t$.

$$L_t u = F_1 + \beta w - \Upsilon v$$

$$L_t v = F_2 + \Upsilon u - \alpha w$$

$$L_t w = F_3 + \alpha v - \beta u$$

Operating with the appropriate inverse of each operator, we have

$$u = u_0 - L_t^{-1} \beta w + L_t^{-1} \Upsilon v$$
$$v = v_0 - L_t^{-1} \Upsilon u + L_t^{-1} \alpha w$$
$$w = w_0 - L_t^{-1} \alpha v + L_t^{-1} \beta u$$

where

$$u_0 = u(\overline{x},0) + L_t^{-1} F_1$$

$$v_0 = v(\overline{x},0) + L_t^{-1} F_2$$

$$w_0 = w(\overline{x},0) + L_t^{-1} F_3$$

Equivalently,

$$u_0 = \Phi(\overline{x}) + L_t^{-1} F_1$$

$$v_0 = \sigma(\overline{x}) + L_t^{-1} F_2$$

$$w_0 = \zeta(\overline{x}) + L_t^{-1} F_3$$

where $\phi(\overline{x})$, $\sigma(\overline{x})$, $\zeta(\overline{x})$ satisfy respectively $Lu = 0$, $Lv = 0$, $Lw = 0$ with appropriate initial or boundary conditions. Making the decompositions $u = \sum_{n=0}^{\infty} u_n$, $v = \sum_{n=0}^{\infty} v_n$, $w = \sum_{n=0}^{\infty} w_n$, we now have

$$\sum_{n=0}^{\infty} u_n = u_0 - L_t^{-1} \beta \sum_{n=0}^{\infty} w_n + L_t^{-1} \Upsilon \sum_{n=0}^{\infty} v_n$$

$$\sum_{n=0}^{\infty} v_n = v_0 - L_t^{-1} \Upsilon \sum_{n=0}^{\infty} u_n + L_t^{-1} \propto \sum_{n=0}^{\infty} w_n$$

$$\sum_{n=0}^{\infty} w_n = w_0 - L_t^{-1} \propto \sum_{n=0}^{\infty} v_n + L_t^{-1} \beta \sum_{n=0}^{\infty} u_n$$

Now the following components are calculable:

$$u_1 = - L_t^{-1} \beta w_0 + L_t^{-1} \Upsilon v_0$$

$$v_1 = - L_t^{-1} \Upsilon u_0 + L_t^{-1} \propto w_0$$

$$w_1 = - L_t^{-1} \propto v_0 + L_t^{-1} \beta u_0$$

Thus, knowing u_0, v_0, w_0, we can determine u_1, v_1, w_1. Similarly we find u_2, v_2, w_2 in terms of u_1, v_1, w_1 etc., so that

$$u_{n+1} = - L_t^{-1} \beta w_n + L_t^{-1} \Upsilon v_n$$

$$v_{n+1} = - L_t^{-1} \Upsilon u_n + L_t^{-1} \propto w_n$$

$$w_{n+1} = - L_t^{-1} \propto v_n + L_t^{-1} \beta u_n$$

for $n > 1$. Components of u, v, w can now be determined to a value of n which yields some sufficient accuracy.

2) Returning to equation (18.4) and assuming $\rho = \mu = 1$, $\nabla p = 0$ and neglecting the constant term, we have

$$\partial u/\partial t + [u\, \partial u/\partial x + v\, \partial u/\partial y + w\, \partial u/\partial z] + 2[\beta w - \Upsilon v]$$

$$= F_1 + [d^2u/\partial x^2 + \partial^2 u/\partial y^2 + \partial^2 u/\partial z^2]$$

$$\partial v/\partial t + [u\, \partial v/\partial x + v\, \partial v/\partial y + w\, \partial v/\partial z] + 2[\Upsilon u - \alpha w]$$

$$= F_2 + [d^2v/\partial x^2 + \partial^2 v/\partial y^2 + \partial^2 v/\partial z^2]$$

$$\partial w/\partial t + [u\, \partial w/\partial x + v\, \partial w/\partial y + w\, \partial w/\partial z] + 2[\alpha v - \beta u]$$

$$= F_3 + [\partial^2 w/\partial x^2 + \partial^2 w/\partial y^2 + \partial^2 w/\partial z^2]$$

We now rearrange these as follows:

$$\partial u/\partial t + \partial^2 u/\partial x^2 + \partial^2 u/\partial y^2 + \partial^2 u/\partial z^2 = F_1 - 2[\beta w - \Upsilon v]$$

$$- [u\, \partial u/\partial x + v\, \partial u/\partial y + w\, \partial u/\partial z]$$

$$\partial v/\partial t + \partial^2 v/\partial x^2 + \partial^2 v/\partial y^2 + \partial^2 v/\partial z^2 = F_2 - 2[\Upsilon u - \alpha w]$$

$$- [u\, \partial v/\partial x + v\, \partial v/\partial y + w\, \partial v/\partial z]$$

$$\partial w/\partial t + \partial^2 w/\partial x^2 + \partial^2 w/\partial y^2 + \partial^2 w/\partial z^2 = F_3 - 2[\alpha v - \beta u]$$

$$- [u\, \partial w/\partial x + v\, \partial w/\partial y + w\, \partial w/\partial z]$$

Defining $L_t = \partial/\partial t$, $L_x = \partial^2/\partial x^2$, $L_y = \partial^2/\partial y^2$, $L_z = \partial^2/\partial z^2$ with L_t^{-1} defined accordingly as the definite integral from 0 to t, L_x^{-1}, L_y^{-1}, L_z^{-1} as the two-fold indefinite integrations , respectively for initial conditions specified, we have

$$[L_t + L_x + L_y + L_z]u = F_1 - R_1(u,v,w) - N_1(u,v,w)$$

$$[L_t + L_x + L_y + L_z]v = F_2 - R_2(u,v,w) - N_2(u,v,w)$$

$$[L_t + L_x + L_y + L_z]w = F_3 - R_3(u,v,w) - N_3(u,v,w)$$

where $R_1 = 2(\beta w - \Upsilon v)$, $R_2 = 2(\Upsilon u - \alpha w)$, $R_3 = 2(\alpha v - \beta u)$, and

$$N_1(u,v,w) = u\ \partial u/\partial x + v\ \partial u/\partial y + w\ \partial u/\partial z$$

$$N_2(u,v,w) = u\ \partial v/\partial x + v\ \partial v/\partial y + w\ \partial v/\partial z$$

$$N_3(u,v,w) = u\ \partial w/\partial x + v\ \partial w/\partial y + w\ \partial w/\partial z$$

For notational purposes we can, if we wish, define a nonlinear operator

$$N[\cdot] = [u\partial/\partial x + v\partial/\partial y + w\partial/\partial z][\cdot]$$

and instead of $N_1(u,v,w)$, $N_2(u,v,w)$, $N_3(u,v,w)$, we can write Nu, Nv, Nw, so long as we remember that each of these depends on u, v, w. For brevity writing $L = L_t + L_x + L_y + L_z$, we have

$$Lu = F_1 - R_1(u,v,w) - Nu$$

$$Lv = F_2 - R_2(u,v,w) - Nv$$

$$Lw = F_3 - R_3(u,v,w) - Nw$$

Consider the first of these equations and solve each component of Lu thus.

$$L_t u = F_1 - R_1(u,v,w) - Nu - L_x u - L_y u - L_z u$$

$$L_x u = F_1 - R_1(u,v,w) - Nu - L_y u - L_z u - L_t u$$

$$L_y u = F_1 - R_1(u,v,w) - Nu - L_z u - L_t u - L_x u$$

$$L_z u = F_1 - R_1(u,v,w) - Nu - L_t u - L_x u - L_y u$$

Operating with the appropriate inverses L_t^{-1} on the first equation, L_x^{-1} on the second, L_y^{-1} on the third, and L_z^{-1} on the fourth, we can write:

$$u = \phi_1 + L_t^{-1} F_1 - L_t^{-1} R_1(u,v,w) - L_t^{-1} Nu$$

$$- [L_t^{-1} L_x + L_t^{-1} L_y + L_t^{-1} L_z]u$$

$$u = \phi_2 + L_x^{-1} F_1 - L_x^{-1} R_1(u,v,w) - L_x^{-1} Nu$$
$$- [L_x^{-1} L_y + L_x^{-1} L_z + L_x^{-1} L_t]u$$

$$u = \phi_3 + L_y^{-1} F_1 - L_y^{-1} R_1(u,v,w) - L_y^{-1} Nu$$
$$- [L_y^{-1} L_z + L_y^{-1} L_t + L_y^{-1} L_x]u$$

$$u = \phi_4 + L_z^{-1} F_1 - L_z^{-1} R_1(u,v,w) - L_z^{-1} Nu$$
$$- [L_z^{-1} L_t + L_z^{-1} L_x + L_z^{-1} L_y]u$$

where ϕ_1, ϕ_2, ϕ_3, ϕ_4 are solutions, respectively, of $L_t u = 0$, $L_x u = 0$, $L_y u = 0$, $L_z u = 0$, and constants can be evaluated from boundary conditions. In the case where only initial conditions are specified, we will have

$$\phi_1 = u(x,y,z,0)$$

$$\phi_2 = u(0,y,z,t) + xu_x(0,y,z,t)$$

$$\phi_3 = u(x,0,z,t) + yu_y(x,0,z,t)$$

$$\phi_4 = u(x,y,0,t) + zu_z(x,y,0,t)$$

Now we add the four equations and divide by four. Letting

$$u_0 = (1/4)(\phi_1 + \phi_2 + \phi_3 + \phi_4) + [L_t^{-1} + L_x^{-1} + L_y^{-1} + L_z^{-1}] F_1$$

we have

$$u = u_0 - (1/4)\{[L_t^{-1} + L_x^{-1} + L_y^{-1} + L_z^{-1}]R_1(u,v,w)\}$$
$$- (1/4)\{[L_t^{-1} + L_x^{-1} + L_y^{-1} + L_z^{-1}] N(u)\}$$

$$- (1/4)\{[L_t^{-1} L_x + L_t^{-1} L_y + L_t^{-1} L_z]$$

$$+ [L_x^{-1} L_y + L_x^{-1} L_z + L_x^{-1} L_t]$$

$$+ [L_y^{-1} L_z + L_y^{-1} L_t + L_y^{-1} L_x]$$

$$+ [L_z^{-1} L_t + L_z^{-1} L_x + L_z^{-1} L_y]\}u$$

Now let $u = \sum_{n=0}^{\infty} u_n$, $v = \sum_{n=0}^{\infty} v_n$, $w = \sum_{n=0}^{\infty} w_n$ and let $N(u) = \sum_{n=0}^{\infty} A_n$, where the A_n are generated for the specific nonlinearity. Finally, we do the same with the equations for v and w. As discussed above we already have u_0, and we find v_0 and w_0 in the same wau. Followingcomponents are found in terms of successive triads, e.g., (u_1,v_1,w_1) in terms of (u_0,v_0,w_0) or $(u_{n+1}, v_{n+1},w_{n+1})$ in terms of (u_n,v_n,w_n) for $n \geq 0$.

If we write $L_t^{-1} + L_x^{-1} + L_y^{-1} + L_z^{-1} = L_1$ and $L_t^{-1} [L_x + L_y + L_z] + L_x^{-1} [L_y + L_z + L_t] + L_y^{-1} [L_z + L_t + L_x] + L_z^{-1} [L_t + L_x + L_y] = L_2$, we have

$$u = u_0 - (1/4)L_1^{-1} [\beta \sum_{n=0}^{\infty} w_n - \Upsilon \sum_{n=0}^{\infty} v_n]$$

$$- (1/4)L_1^{-1} \sum_{n=0}^{\infty} A_n - (1/4)L_2 \sum_{n=0}^{\infty} u_n$$

$$v = v_0 - (1/4)L_1^{-1} [\Upsilon \sum_{n=0}^{\infty} u_n - \alpha \sum_{n=0}^{\infty} w_n]$$

$$- (1/4)L_1^{-1} \sum_{n=0}^{\infty} B_n - (1/4)L_2 \sum_{n=0}^{\infty} v_n$$

$$w = w_0 - (1/4)L_1^{-1} [\alpha \sum_{n=0}^{\infty} v_n - \beta \sum_{n=0}^{\infty} u_n]$$

$$- (1/4)L_1^{-1} \sum_{n=0}^{\infty} C_n - (1/4)L_2 \sum_{n=0}^{\infty} w_n$$

where the A_n, B_n, C_n represent our polynomials for N_1, N_2, N_3 respectively. Now

$$u_{n+1} = - (1/4)L_1^{-1} [\beta w_n - \Upsilon v_n]$$

$$- (1/4)L_1^{-1} A_n - (1/4)L_2 u_n$$

$$v_{n+1} = - (1/4)L_1^{-1} [\Upsilon u_n - \alpha w_n]$$

$$- (1/4)L_1^{-1} B_n - (1/4)L_2 v_n$$

$$w_{n+1} = - (1/4)L_1^{-1} [\alpha v_n - \beta u_n]$$

$$- (1/4)L_1^{-1} C_n - (1/4)L_2 w_n$$

for $n \geq 0$.

We observe that the equations are coupled in both linear and nonlinear terms. The polynomials A_n, B_n, C_n for N_1, N_2, N_3 respectively are found in the same way, and we need discuss only one. Thus

$$Nu = N_1(u,v,w) = \sum_{n=0}^{\infty} A_n = (u\, \partial/\partial x + v\, \partial/\partial y + w\, \partial/\partial z)u$$

Using our previously discussed rule for subscripts,

$$A_0 = u_0\, \partial u_0/\partial x + v_0\, \partial u_0/\partial y + w_0\, \partial u_0/\partial z$$

$$A_1 = u_1\, \partial u_0/\partial x + u_0\, \partial u_1/\partial x + v_1\, \partial u_0/\partial y$$

$$+ v_0\, \partial u_1/\partial y + w_1\, \partial u_0/\partial z + w_0\, \partial u_1/\partial z$$

$$A_2 = u_2\, \partial u_0/\partial x + u_0\, \partial u_2/\partial x + u_1\, \partial u_1/\partial x$$

$$+ v_2\, \partial u_0/\partial y + v_0\, \partial u_2/\partial y + v_1\, \partial u_1/\partial y$$

$$+ w_2\, \partial u_0/\partial z + w_0\, \partial u_2/\partial z + w_1\, \partial u_1/\partial z$$

etc. Alternatively, we can use the analytic parametrization scheme. Let $Nu = \sum_{n=0}^{\infty} A_n \lambda^n$, $u = \sum_{n=0}^{\infty} u_n \lambda^n$, $v = \sum_{n=0}^{\infty} v_n \lambda^n$, $w = \sum_{n=0}^{\infty} w_n \lambda^n$ and identify corresponding terms for $\lambda^0, \lambda^1 \ldots$ which, of course, yields the same results. As discussed in previous work, we caution the reader that λ is not a perturbation parameter, i.e., it is not assumed small. It only is an identifier of terms. The

B_n polynomials simply replace the $u_0, u_1 \ldots,$ with $v_0, v_1, \ldots,$ and the C_n replace the $u_0, u_1, \ldots$ with $w_0, w_1, \ldots$.

We emphasize again that we are solving an equation in the standard form $Lu + Nu = g$. Once it is solved, all equations in this form are solved. It is only necessary to use an invertible part of L; we use the highest ordered derivative, define the inverse as an integral operator, identify the first term properly, and write the nonlinear term in terms of the A_n polynomials. If we have a system of equations, the procedure is similar except that we identify a (vector) first term and proceed as discussed thoroughly in (1986).

Multidimensional Inverse Operator: Consider the equation

$$Lu = F_1$$

where $L = L_t + L_x + L_y + L_z$. Solving for each linear operator term $L_t u$, $L_x u$, ..., in turn, we have

$$L_t u = F_1 - [L_x + L_y + L_z]u$$

$$L_x u = F_1 - [L_y + L_z + L_t]u$$

$$L_y u = F_1 - [L_z + L_t + L_x]u$$

$$L_z u = F_1 - [L_t + L_x + L_y]u$$

Let us suppose that the initial condition terms vanish (only to show an interesting result, not generally true, of course) i.e., the solutions of $L_t u = 0$, $L_x u = 0$, etc., vanish. Then

$$u = L_t^{-1} F_1 - L_t^{-1} [L_x + L_y + L_z]u$$

$$u = L_x^{-1} F_1 - L_x^{-1} [L_y + L_z + L_t]u$$

$$u = L_y^{-1} F_1 - L_y^{-1} [L_z + L_t + L_x]u$$

$$u = L_z^{-1} F_1 - L_z^{-1} [L_t + L_x + L_y]u$$

Adding and dividing by four

$$u = (1/4)[L_t^{-1} + L_x^{-1} + L_y^{-1} + L_z^{-1}]F_1$$
$$- (1/4)\{L_t^{-1}[L_x + L_y + L_z] + L_x^{-1}[L_y + L_z + L_t]$$
$$+ L_y^{-1}[L_z + L_t + L_x] + L_z^{-1}[L_t + L_x + L_y]\}u$$

Let

$$u_0 = (1/4)[L_t^{-1} + L_x^{-1} + L_y^{-1} + L_z^{-1}]F_1$$
$$u_1 = - (1/4)\{L_t^{-1}[L_x + L_y + L_z] + L_x^{-1}[L_y + L_z + L_t]$$
$$+ L_y^{-1}[L_z + L_t + L_x] + L_z^{-1}[L_t + L_x + L_y]\}u_0$$

.

.

.

$$u_{n+1} = - (1/4)\{L_t^{-1}[L_x + L_y + L_z] + L_x^{-1}[L_y + L_z + L_t]$$
$$+ L_y^{-1}[L_z + L_t + L_x] + L_z^{-1}[L_t + L_x + L_y]\}u_n$$

Thus we can write using simply $\{\cdot\}$ for the bracketed quantity above

$$u_1 = - (1/4)\{\cdot\}(1/4)[L_t^{-1} + L_x^{-1} + L_y^{-1} + L_z^{-1}]F_1$$

.

.

$$u_{n+1} = (-1)^n(1/4)^{n+1}\{\cdot\}[L_t^{-1} + L_x^{-1} + L_y^{-1} + L_z^{-1}]F_1$$

$$u = \sum_{n=0}^{\infty} u_n = \sum_{n=0}^{\infty}(-1)^n(1/4)^{n+1}\{(L_x^{-1}L_y + L_y^{-1}L_x)$$
$$+ (L_x^{-1}L_z + L_z^{-1}L_x) + (L_x^{-1}L_t + L_t^{-1}L_x)$$
$$+ (L_y^{-1}L_z + L_z^{-1}L_y) + (L_t^{-1}L_y + L_y^{-1}L_t)$$

$$+ (L_t^{-1} L_z + L_z^{-1} L_t)\}^n \cdot [L_t^{-1} + L_z^{-1} + L_y^{-1} + L_x^{-1}]F_1$$

so that the inverse $L^{-1} = [L_t + L_x + L_y + L_z]^{-1}$ has been identified as:

$$L^{-1} = \sum_{n=0}^{\infty} (-1)^n (1/4)^{n+1} \{(L_x^{-1} L_y + L_y^{-1} L_x)$$

$$+ (L_x^{-1} L_z + L_z^{-1} L_x) + (L_x^{-1} L_t + L_t^{-1} L_x)$$

$$+ (L_y^{-1} L_z + L_z^{-1} L_y) + (L_t^{-1} L_y + L_y^{-1} L_t)$$

$$+ (L_t^{-1} L_z + L_z^{-1} L_t)\}^n \cdot [L_t^{-1} + L_z^{-1} + L_y^{-1} + L_x^{-1}]$$

Now we can return to the equations (18.4)

$$Lu + R_1(u,v,w) + N_1(u,v,w) = g_1$$

$$Lv + R_2(u,v,w) + N_2(u,v,w) = g_2$$

$$Lw + R_3(u,v,w) + N_3(u,v,w) = g_3$$

and solve in terms of the multidimensional operator L rather than the component operators.

$$Lu = g_1 - R_1(u,v,w) - N_1(u,v,w)$$

$$Lv = g_2 - R_2(u,v,w) - N_2(u,v,w)$$

$$Lw = g_3 - R_3(u,v,w) - N_3(u,v,w)$$

One suspects that convergence may be more rapid with this procedure, but it appears easier to use the previous method and simply compute each component in terms of the previous one. We will define

$$u_0 = \phi_u + L^{-1} g_1$$

$$v_0 = \phi_v + L^{-1} g_2$$

$$w_0 = \phi_w + L^{-1} g_3$$

where ϕ_u, ϕ_v, ϕ_w are the homogeneous solutions (see 1986).

$$u = u_0 - L^{-1}R_1(u,v,w) - L^{-1}N_1(u,v,w)$$

$$v = v_0 - L^{-1}R_2(u,v,w) - L^{-1}N_2(u,v,w)$$

$$w = w_0 - L^{-1}R_3(u,v,w) - L^{-1}N_3(u,v,w)$$

$$u_1 = - L^{-1}(2(\beta w_0 - \Upsilon v_0)) - L^{-1}A_0(u_0,v_0,w_0)$$

$$v_1 = - L^{-1}(2(\Upsilon u_0 - \alpha w_0)) - L^{-1}B_0(u_0,v_0,w_0)$$

$$w_1 = - L^{-1}(2(\alpha v_0 - \beta u_0)) - L^{-1}C_0(u_0,v_0,w_0)$$

$$u_2 = - L^{-1}(2(\beta w_1 - \Upsilon v_1)) - L^{-1}A_1(u_0,u_1;v_0,v_1;w_0,w_1)$$

$$v_2 = - L^{-1}(2(\Upsilon u_1 - \alpha w_1)) - L^{-1}B_1(u_0,u_1;v_0,v_1;w_0,w_1)$$

$$w_2 = - L^{-1}(2(\alpha v_1 - \beta u_1)) - L^{-1}C_1(u_0,u_1;v_0,v_1;w_0,w_1)$$

$$\vdots$$

$$u_n = - L^{-1}(2(\beta w_{n-1} - \Upsilon v_{n-1})) - L^{-1}A_{n-1}$$

$$v_n = - L^{-1}(2(\Upsilon u_{n-1} - \alpha w_{n-1})) - L^{-1}B_{n-1}$$

$$w_n = - L^{-1}(2(\alpha v_{n-1} - \beta u_{n-1})) - L^{-1}C_{n-1}$$

The A_n, B_n, C_n are evaluated previously. Returning to (18.5)

$$L_t u + L_x u + L_y u + L_z u = g_1 - R_1(u,v,w) - N_1(u,v,w) \tag{18.6}$$

$$L_t v + L_x v + L_y v + L_z v = g_2 - R_2(u,v,w) - N_2(u,v,w) \tag{18.7}$$

$$L_t w + L_x w + L_y w + L_z w = g_3 - R_3(u,v,w) - N_3(u,v,w) \tag{18.8}$$

From (18.6)

$$L_t u = g_1 - R_1 - N_1 - L_x u - L_y u - L_z u$$

$$L_x u = g_1 - R_1 - N_1 - L_t u - L_y u - L_z u \tag{18.9}$$

$$L_y u = g_1 - R_1 - N_1 - L_x u - L_t u - L_z u$$

$$L_z u = g_1 - R_1 - N_1 - L_x u - L_y u - L_t u$$

From (18.7)

$$L_t v = g_2 - R_2 - N_2 - L_x v - L_y v - L_z v$$

$$L_x v = g_2 - R_2 - N_2 - L_t v - L_y v - L_z v \tag{18.10}$$

$$L_y v = g_2 - R_2 - N_2 - L_x v - L_t v - L_z v$$

$$L_z v = g_2 - R_2 - N_2 - L_x v - L_y v - L_t v$$

From (18.8)

$$L_t w = g_3 - R_3 - N_3 - L_x w - L_y w - L_z w$$

$$L_x w = g_3 - R_3 - N_3 - L_t w - L_y w - L_z w$$

$$L_y w = g_3 - R_3 - N_3 - L_x w - L_t w - L_z w \tag{18.11}$$

$$L_z w = g_3 - R_3 - N_3 - L_x w - L_y w - L_t w$$

The required inversions yield the auxiliary conditions whose functional form must be specified in any particular flow problems. Thus with the required inversions, addition of the equations for each component, and dividing by four (18.9) becomes

$$\begin{aligned} u = u_0 - (1/4)\{&[L_t^{-1} + L_x^{-1} + L_y^{-1} + L_z^{-1}]R_1 \\ &+ [L_t^{-1} + L_x^{-1} + L_y^{-1} + L_z^{-1}]N_1 \\ &+ [(L_t^{-1} L_x + L_t^{-1} L_y + L_t^{-1} L_z) \\ &+ (L_x^{-1} L_t + L_x^{-1} L_y + L_x^{-1} L_z) \end{aligned}$$

$$+ (L_y^{-1} L_x + L_y^{-1} L_t + L_y^{-1} L_z)$$

$$+ (L_z^{-1} L_x + L_z^{-1} L_y + L_z^{-1} L_t)]\}\sum_{n=0}^{\infty} u_n$$

where u_0 contains the terms involving g_1 as well as the auxiliary conditions from the inversions.

For simplicity let $K = L_t^{-1} + L_x^{-1} + L_y^{-1} + L_z^{-1}$ and let G represent the remaining terms operating on $u = \sum_{n=0}^{\infty} u_n$. Thus,

$$u = u_0 - (1/4)\{KR_1 + KN_1 + G\} \sum_{n=0}^{\infty} u_n$$

Similarly,

$$v = v_0 - (1/4)\{KR_2 + KN_2 + G\} \sum_{n=0}^{\infty} v_n$$

$$w = w_0 - (1/4)\{KR_3 + KN_3 + G\} \sum_{n=0}^{\infty} w_n$$

Thus if we know u_0, v_0, w_0, then all other terms are calculable, each term from the preceding. It is not, perhaps, obvious from the above form written for simplicity, but the equations are certainly coupled since the R's and N's depend on all the components. Thus we determine the triad u_0, v_0, w_0, then the triad u_1, v_1, w_1

$$u_1 = -(1/4)\{KR_1(u_0,v_0,w_0) + KN_1(u_0,v_0,w_0) + G\}u_0$$

$$v_1 = -(1/4)\{KR_2(u_0,v_0,w_0) + KN_2(u_0,v_0,w_0) + G\}v_0$$

$$w_1 = -(1/4)\{KR_3(u_0,v_0,w_0) + KN_3(u_0,v_0,w_0) + G\}w_0$$

etc., to find all decomposition components of the velocity components u, v, w. We must now specify u_0, v_0, w_0, which include the terms arising from the inversion and the forcing functions. These will also involve a one-fourth factor from the additions, e.g., of the terms of inversions in (18.9). Since $L_t = \rho\, \partial/\partial t$,

$$L_t^{-1} L_t u = 1/\rho \int_0^t (\rho\, du/dt)dt$$

If ρ were constant, we would have $u(x,y,z,t) - u(x,y,z,0)$ on the left, and, of course, $L_t^{-1} g_1$. The $u(x,y,z,0) + L_t^{-1} g$ would be added to similar terms of the following equations (which involve second derivatives and yield two initial/boundary conditions each) and divided by four to get u_0. Of course, in the equations involving inversions of L_x, L_y, L_z, each yielding two conditions, we also have the μ, which is not necessarily constant, to complicate the otherwise trivial inversions. To proceed in this (deterministic) version, we must specify the density and viscosity and carry out the above integrations. Then specification of initial/boundary conditions and the input g allows determination of u_0, v_0, w_0 from which all other components are now determinable.

Stochastic Model: The intrinsic nonlinearity has not been the only difficulty precluding direct analytic approach. An essential stochastic nature has also precluded understanding of phenomena such as turbulence. Of course, we are also proposing approximation, but not the commonly used approximations which change the problem to a tractable one, or which saturate users with printouts and computation and lose sight of analytical dependences. We propose solution of the nonlinear stochastic problem without assumptions of weak nonlinearity or small fluctuations or discretization. If the result solves the real physical problem, converges, is verifiable, then a unique opportunity for advances in our understanding exists. The solution earlier of the Burger's equation provided one test of our analytic approximation (or decomposition) method since it can be solved explicitly by a change of variable. Calculations by standard methods, such as finite differences, are constrained by computing time and required storage with computational time sometimes rising to unmanageable limits.

Suppose we consider the equation when stochasticity is involved; we then have:

$$Lu = F_1 - L_1 - Nu$$

$$Lv = F_2 - L_2 - Nv$$

$$Lw = F_3 - L_3 - Nw$$

where we have defined $N \equiv u\, \partial/\partial x + v\, \partial/\partial y + w\, \partial/\partial z$ and L, L_1, L_2,

L_3 are defined as before. Solving as before we write u, v, and w in decomposition form to determine their components. Of course, we will take some finite approximations ϕ_u, ϕ_v, ϕ_w for u, v, and w. The quantities ϕ_u, ϕ_v, ϕ_w are given as a (finite) series of stochastic terms since the velocity $\vec{U}$ or (u,v,w) may be random and the parameters in the equation may be random. Averaging ϕ_u, ϕ_v, ϕ_w we have the expected velocities $\langle\phi_u\rangle$, $\langle\phi_v\rangle$, $\langle\phi_w\rangle$, or $\langle u\rangle$, $\langle v\rangle$, $\langle w\rangle$, remembering that they are n-term approximations which we can get as accurately as we like. The second order statistics then are found as in (1983), e.g., the covariance function for u by considering the product at two time instants of the quantity $u - \langle u\rangle$ and averaging the result.

We can now look forward to solving strongly nonlinear, strongly stochastic systems of equations, even involving delay terms and position and time variables x, y, z, t with very complicated boundary conditions (integral boundary conditions, coupled boundary conditions, etc.). However, it is important to realize that the Navier-Stokes equation is undoubtedly incorrect in the stochastic case and requires remodeling from the beginning. The deterministic case is a limiting case of the stochastic model; one cannot simply replace p, ρ, μ, ν by stochastic equivalents, just as in the case of a wave propagating in a random medium, we cannot simply replace the velocity in the D'Alembertian with a stochastic quantity. Until current research on better models is completed, after which quantitative comparisons are in order, we must proceed with the current model.

We write the velocity $u(x,y,z,t) = U(x,y,z,t) + u'(x,y,z,t)$ where $\langle u(x,y,z,t)\rangle$ is an expected or time-averaged flow and u' involves random fluctuations from the mean. Now, all other stochastic quantities such as ρ, p, etc., are handled the same way. Products of mean and primed quantities are zero, but products of prime quantities, when averaged, can be very large and essentially couple the large-scale motions and small-scale motions. Large-scale motions on the rotating earth must take into account Coriolis force. If we consider only small-scale motions or a small part of the ocean, we can neglect this force.

Looking at the continuity equation we see that if $d\rho/dt = 0$, i.e., if the fluid is incompressible, then the divergence of the vector field vanishes. This is a customary assumption in the derivations, so we see the div $\rho u = 0$ instead of the hydrodynamic derivative $\nabla \cdot \rho u + \partial\rho/\partial t = 0$. If the partial derivative of ρ with respect to time is zero, then $\nabla \cdot u = 0$ (source-free field). (Vanishing divergence means $u = \nabla\Psi$ where Ψ is a potential function. Lines of Ψ = constant are called streamlines. $\nabla \cdot u = 0$ says that for a *horizontal* incompressible current field, $\partial u/\partial x + \partial v/\partial y = 0$ where u, v, w are components of u. Hence $u = \partial\Psi/\partial y$ and $v = -\partial\Psi/\partial x$. To get lines of constant Ψ, we set $d\Psi = 0$ or $(\partial\Psi/\partial x)dx + (\partial\Psi/\partial y)dy = 0$; hence the inclination of the constant Ψ lines $(dy/dx)_{\phi=const} = v/u$ which is the angle between current direction and the x axis.)

A real ocean, of course, would not have irrotational or curl-free motion since frictional forces are involved. Curl is also called vorticity in hydrodynamics. In the curl-free, or vorticity-free, case for the current we have curl $\vec{u} = 0$ hence $\vec{u} = \nabla\phi$, where ϕ is called the velocity potential. (In the special case of a horizontal xy plane $\partial v/\partial x = \partial u/\partial y = 0$, so $u = \partial\phi/\partial x$ and $v = \partial\phi/\partial y$.) If we have horizontal flow with a vertical velocity profile in which the horizontal velocity component $u(z)$ varies with z, the velocity at one level of a water particle is different from the velocity of a water particle below or above it, so there will be a circulation in a vertical plane and flow will not be curl-free.

For a real ocean, quantities such as velocity, density, pressure, temperature can vary from point to point and may manifest violent fluctuations in some circumstances. The flow in the ocean (or atmosphere) is not laminar; it involves an irregular motion with rapid complex fluctuations in velocity which constitute "turbulence." Treating these fluctuations as quantities to be averaged out to explain gross features is too simplistic, and a full stochastic treatment is called for. Thus, we will let u, v w be components of the velocity $\vec{u}$ and write $u = \langle u \rangle + u'$, $v = \langle v \rangle + v'$, $w = \langle w \rangle + w'$, and let u', v', w' represent the turbulent (random) so-called "small scale" fluctuations around the average state.

Turbulent flow (at fairly large Reynolds numbers), also called fully developed turbulence, is characterized by the presence of extremely random variations of velocity with respect to time and between points in the flow at a given instant. The amplitude of the fluctuations is generally *not* small with respect to the mean value, so perturbative treatments will not explain turbulence. The mathematical treatments of turbulence have been completely inadequate because turbulence, such as we are discussing, is a strongly stochastic, strongly nonlinear phenomena and cannot be dealt with and understood with linearized models and "small" fluctuations, closure approximations, and perturbation theories.

For finding averages from our hydrodynamical equations, we will avoid the usual hierarchy methods, since the required closure approximations are a limitation to small fluctuations just as in perturbation theories. The classical hydrodynamical equations including these turbulent or fluctuating terms are said to be valid for the average, *if* certain terms are added to the average flow ("eddy functional effects"). This is an unrealistic mathematical artifice which neglects some possible fluctuations in density and pressure and is no substitute for a correct stochastic solution. Yet much of the so-called theory of turbulence is concerned with the approximation of these added "Reynolds stresses" in the hope that certain properties of the (high wave number) spectra can be found from the *mean* motion. Turbulence can be thought of as that part of the fluid motion caused by the stochasticity of velocity u, pressure p, density ρ, and possibly viscosity and salinity. Thus we must solve very complex coupled forms of the nonlinear stochastic operator equations those we consider in (1983; 1986) in which differential operators involve stochastic process coefficients. From these stochastic equations we need to calculate first-and second-order statistical measures: then, we will have determined mean behavior and variances of the turbulence.

A close connection clearly exists between turbulence and the generation and decay of vortices and internal waves and propagation in a random media since internal waves below the sea surface generate velocity fields as do collapsing turbulent wakes and disturbances due to undersea transit.

A sequence of numbers representing observation of a quantity of interest at equally spaced time intervals is a realization of a so-called "time series." Such series occur obviously in meteorological or oceanographic, economic and business, or physical observations. The underlying process is a stochastic process, and the result of any continuous observation is a realization of that process or a sample function . The discrete uniformly spaced observations simply yield the functional values of the realization at those instants of time or discrete numbers (or series of numbers at each time if it is a vector stochastic process). In geophysics, particularly oceanography and turbulence, these processes are non-stationary processes - not to be studied by power spectra, for example; however, for simplicity, assumptions are often made that the statistically changing nature occurs sufficiently slowly that stationarity is an adequate assumption. The Gaussian assumption is also standard - again, for mathematical rather than physical reasons and is not actually necessary to us. We need say nothing about probability density functions but only about statistical measures - expectation and two-point covariance functions. (Probability densities are discussed in 1983.) We will proceed then in our consideration of the Navier-Stokes equations.

The vector equation can be written as a system of coupled nonlinear stochastic equations and solved as a coupled system as before. The A_n for $N\overline{u}$ and $\mathcal{M}\overline{u}$ are found in the same manner. With the use of the decomposition method, the resulting series involves stochastic processes and using ϕ_n as the sum of n terms of the series where n is the number of terms which provide a sufficiently accurate solution, we obtain expectation and two-point covariance from ϕ_n.

Proceeding then with the stochastic case which, as stated, should be remodeled, we write density as $\rho = \langle\rho\rangle + \rho_r$ where ρ is the expected value and ρ_r is the fluctuating component. Similarly viscosity $\mu = \langle\mu\rangle + \mu_r$ and $p = \langle p\rangle + p_r$ (we will write ρ or μ, of course, for the entire quantity). Defining now

$$L\overline{u} = \langle\rho\rangle\, \partial\overline{u}/\partial t$$

$$R\overline{u} = -\langle\mu\rangle\, \nabla^2\overline{u} + (\mu/3)\, \nabla\, (1/\rho)(\overline{u}\cdot\nabla)\rho$$

$$\mathcal{R}\overline{u} = \rho_r\, \partial\overline{u}/\partial t - \mu_r\, \nabla^2\, \overline{u}$$

$$N\overline{u} = \langle\rho\rangle(\overline{u}\cdot\nabla)\overline{u} + 2\langle\rho\rangle(\Omega_e \times \overline{u})$$

$$\mathcal{M}\overline{u} = \rho_r(\overline{u}\cdot\nabla)\overline{u} + 2\rho_r(\Omega_e \times \overline{u})$$

$$g = \overline{F} - (\mu/3)\, \nabla\, (1/\rho)(\partial\rho/\partial t) - \nabla p - kg\rho$$

which is in our usual form $\mathcal{L}\overline{u} + \mathcal{N}\overline{u} = g$, with $\mathcal{L}\overline{u}$ a (stochastic) linear term and $\mathcal{N}\overline{u}$ a (stochastic) nonlinear term. The term $\mathcal{L}\overline{u} = L\overline{u} + R\overline{u} + \mathcal{R}\overline{u}$ where L is a linear deterministic invertible operator, R is the remainder (linear deterministic) operator, $\mathcal{R}$ is a linear stochastic component. Similarly $\mathcal{N}\overline{u} = N\overline{u} + \mathcal{M}\overline{u}$, with N deterministic and $\mathcal{M}$ stochastic. At this point, it is clear we can solve the equation since the equation $\mathcal{F}u = \mathcal{L}u + \mathcal{N}u = g$ has previously been solved.

We will avoid many commonly used approximations, e.g., the Boussinesq approximation which neglects significant stochastic effects by dropping most (but not all) terms involving the random part of ρ. Though validity is claimed, the fluctuations in the coefficients of a differential equation *do* affect the solution process. The decomposition method yields statistical separability for such approximations naturally which does not occur with the ordinary averaging methods used.

References

Adomian, G., *Nonlinear Stochastic Operator Equations* , Academic Press, 1986.

Adomian, G., *Stochastic Systems* , Academic Press 1983.

Adomian, G., "Systems of Nonlinear Partial Differential Equations," *J. Math. Anal. and Applic.* , 115, 1, 1986, 235-238.

Bellman, R. E. and G. Adomian. *Partial Differential Equations - New Methods for Their Treatment and Application* , Reidel, 1986.

Bellomo, N. and R. Riganti, *Nonlinear Stochastic Systems Theory in Physics and Mechanics* , World Scientific Publishing Company, 1987.

Epilogue

Some previous discussions of convergence have not been entirely satisfactory from a rigorous viewpoint so legitimate questions have been raised by Hazewinkel and others about it. Let us take a final look. Consider the differential equation $Lu + Ru + Nu = g$ where $L + R$ represents the total linear operator with L an invertible operator - usually the highest-ordered differential, R a "remainder" operator and Nu a nonlinear term. Assume $Nu = f(u)$, an analytic function. We will assume g and coefficients in the equation appropriately smooth and bounded. We are dealing with *physical* systems in which solutions exist. The decomposition solution is given by $\sum_{n=0}^{\infty} u_n$ where

$$u_0 = \phi + L^{-1}g$$

where $L\phi = 0$ and

$$u_{n+1} = - L^{-1}Ru_n - L^{-1}A_n$$

for $n \geq 0$ where the A_n are defined on page 13. Substituting the above "solution" in the differential equation,

$$Lu = L \sum_{n=0}^{\infty} u_n$$

$$= L[\phi + L^{-1}g] - L[L^{-1}R \sum_{n=0}^{\infty} u_n] - L[L^{-1} \sum_{n=0}^{\infty} A_n]$$

$$= g - Ru - \sum_{n=0}^{\infty} A_n$$

or

$$Lu + Ru + \sum_{n=0}^{\infty} A_n = g$$

If $f(u) = \sum_{n=0}^{\infty} A_n$, the equation is satisfied.

For the A_n, we have (see page 13)

$$A_0 = f(u_0)$$

$$A_1 = u_1 df/du_0$$

$$A_2 = u_2 df/du_0 + (1/2)u_1^2 d^2 f/du_0^2$$

$$A_3 = u_3 df/du_0 + u_1 u_2 d^2 f/du_0^2 + (1/6)u_1^3 d^3 f/du_0^3$$

.

.

.

If we write $\sum_{n=0}^{\infty} A_n$ and rearrange the terms, we have exactly

$$\sum_{n=0}^{\infty} A_n = f(u_0) + (u - u_0)\, df/du_0 + [(u - u_0)/2]\, d^2 f/du_0^2 + \cdots$$

or a generalized Taylor series for $f(u)$ about the function u_0 rather than about a point. Thus $fu = \sum_{n=0}^{\infty} A_n$ and the decomposition solution is the solution.

If we begin with a partial differential equation

$$L_t u + L_x u = g(x,t)$$

considering only two dimensions for simplicity, the decomposition solution is $\sum_{n=0}^{\infty} u_n$ where

$$u_0 = (1/2)(\phi_t + \phi_x)$$

(where ϕ_t satisfies $L_t\phi_t = 0$ and ϕ_x satisfies $L_x\phi_x = 0$) and

$$u_1 = -(1/2)[L_t^{-1} L_x + L_x^{-1} L_t]u_0$$

$$u_2 = -(1/2)[L_t^{-1} L_x + L_x^{-1} L_t]u_1$$

.

.

.

If we operate on $(L_t + L_x)u = g$ with $(L_t^{-1} + L_x^{-1})$, we have

$$u - \phi_t + u - \phi_x + [L_x^{-1} L_t + L_t^{-1} L_x]u$$

$$= (L_t^{-1} + L_x^{-1})g$$

or solving for u

$$u = u_0 - (1/2)[L_x^{-1} L_t + L_t^{-1} L_x]u$$

$$u_0 = (1/2)(\phi_t + \phi_x) + (1/2))(L_t^{-1} + L_x^{-1}]g$$

$$u_1 = -(1/2)[L_x^{-1} L_t + L_t^{-1} L_x]u_0$$

.

.

.

If we begin with $(L_t + L_x)u + f(u) = g$, we proceed in the same way which verifies our solution of $u = \sum_{n=0}^{\infty} u_n$ as assumed in the decomposition method. Equivalently, we can write u as well as $f(u)$ as sums of A_n polynomials generated for the particular function, i.e. $u = \sum_{n=0}^{\infty} A_n\{u\}$ and $f(u) = \sum_{n=0}^{\infty} A_n\{f(u)\}$ since the $A_n\{u\}$ are the u_n.

Index